I0053416

# Wastewater Treatment Using Green Synthesis

*Wastewater Treatment Using Green Synthesis* discusses advances in wastewater treatment with a focus on biological processes. Major topics discussed include bioremediation through microorganisms, green and iron-based nanoparticles in wastewater treatment, $TiO_2$ Doped Lignocellulosic Biopolymer with an emphasis on their photodegradation potential for a variety of pollutants, and permanent film growth. It further includes remediation of industrial sludges/effluents with a particular emphasis on biological treatment, phytoremediation, and de-ballasting water treatment technology.

## FEATURES:

- Focuses on the implementation of effective wastewater treatment with orientation towards biological processes
- Covers both biological and physico-chemical waste and wastewater treatment processes, focusing on emerging techniques
- Reviews computational method approaches, the application of $TiO_2$-doped biopolymers, iron nanoparticles, and the use of molecular techniques for wastewater treatment
- Illustrates molecular docking, molecular dynamics simulation, homology modelling, and biodegradation pathways prediction
- Includes dedicated chapters on biological wastewater treatment

This book is aimed at graduate students and researchers in environmental and chemical engineering with a focus on wastewater treatment.

# Wastewater Treatment Using Green Synthesis

Edited by Swapnila Roy,
Tien Anh Tran,
Zaira Zaman Chowdhury and
Prathibha B.S.

CRC Press
Taylor & Francis Group
Boca Raton London New York

CRC Press is an imprint of the
Taylor & Francis Group, an **informa** business

Designed cover image: © Shutterstock

First edition published 2024
by CRC Press
2385 NW Executive Center Drive, Suite 320, Boca Raton FL 33431

and by CRC Press
4 Park Square, Milton Park, Abingdon, Oxon, OX14 4RN

*CRC Press is an imprint of Taylor & Francis Group, LLC*

© 2024 selection and editorial matter, Swapnila Roy, Tien Anh Tran, Zaira Zaman Chowdhury and Prathibha B.S.; individual chapters, the contributors

Reasonable efforts have been made to publish reliable data and information, but the author and publisher cannot assume responsibility for the validity of all materials or the consequences of their use. The authors and publishers have attempted to trace the copyright holders of all material reproduced in this publication and apologize to copyright holders if permission to publish in this form has not been obtained. If any copyright material has not been acknowledged please write and let us know so we may rectify in any future reprint.

Except as permitted under U.S. Copyright Law, no part of this book may be reprinted, reproduced, transmitted, or utilized in any form by any electronic, mechanical, or other means, now known or hereafter invented, including photocopying, microfilming, and recording, or in any information storage or retrieval system, without written permission from the publishers.

For permission to photocopy or use material electronically from this work, access www.copyright.com or contact the Copyright Clearance Center, Inc. (CCC), 222 Rosewood Drive, Danvers, MA 01923, 978–750–8400. For works that are not available on CCC please contact mpkbookspermissions@tandf.co.uk

Trademark notice: Product or corporate names may be trademarks or registered trademarks and are used only for identification and explanation without intent to infringe.

ISBN: 978-1-032-37966-1 (hbk)
ISBN: 978-1-032-37969-2 (pbk)
ISBN: 978-1-003-34283-0 (ebk)

DOI: 10.1201/9781003342830

Typeset in Times
by Apex CoVantage, LLC

# Contents

Preface..........................................................................................................xi
About the Editors....................................................................................... xiii
List of Contributors...................................................................................xvii


**Chapter 1**   Bioremediation of Toxic Pollutants ........................................1

*Vidya Acharya, Venkatalakshmi Jakka, Arthita Ray,*
*Sai Rupa Mopidevi, Anjali Syamala, Aniruddha*
*Mukhopadhyay and Shubhalakshmi Sengupta*

1.1    Introduction ...........................................................1
       1.1.1   Bioremediation ...........................................1
1.2    Dyes........................................................................3
       1.2.1   Classification of Synthetic Dyes...................3
1.3    Bioremediation of Dyes........................................3
       1.3.1   By the Use of Bacteria..................................4
       1.3.2   By the Use of Fungi......................................4
       1.3.3   By the Use of Algae .....................................7
       1.3.4   By the Use of Yeast .....................................7
       1.3.5   By the Use of Enzymes ................................7
1.4    Heavy Metals.........................................................7
1.5    Heavy Metal Bioremediation............................. 11
       1.5.1   By the Use of Bacteria................................ 11
       1.5.2   By the Use of Fungi.................................... 11
       1.5.3   By the Use of Algae ................................... 11
1.6    Organic Pollutants ............................................. 11
1.7    Bioremediation of Organic Pollutants ............... 12
       1.7.1   By the Use of Bacteria................................ 12
       1.7.2   By the Use of Algae ................................... 13
1.8    Microplastics ...................................................... 13
1.9    Microplastics Bioremediation............................ 13
       1.9.1   By the Use of Bacteria................................ 14
       1.9.2   By the Use of Fungi.................................... 14
       1.9.3   By the Use of Algae ................................... 14
       1.9.4   Role of Enzymes ....................................... 18
1.10   Bioremediation of Pharmaceutical Waste ..........20
1.11   Advantages and Limitations................................ 21
1.12   Conclusion .......................................................... 21
Acknowledgments ............................................................. 21
References ......................................................................... 21

**Chapter 2**  Green Synthesis of Iron Oxide Nanoparticles and Its
              Application in Water Treatment ............................................................28

*Alli Malar Harikrishnan, Zaira Zaman Chowdhury, Masud
Rana, Ahmed Elsayid Ali, Ajita Mitra, Rahman Faizur Rafique
and Rafie Bin Johan*

2.1    Introduction ..................................................................................28
2.2    Various Green Technology for Synthesis of Iron Oxide
       Nanoparticles ................................................................................ 31
       2.2.1    Green Synthesis of IONPs Using Fungal Biomass .... 31
       2.2.2    Green Synthesis of IONPs Using Algal Biomass ....... 33
       2.2.3    Green Synthesis of IONPs Using Bacteria ................. 33
       2.2.4    Green Synthesis of IONPs Using Plant Species ......... 34
2.3    Applications of Iron Oxide Nanoparticles for Water
       Treatment ...................................................................................... 34
       2.3.1    Iron Oxide Nanoparticles as Adsorbent for
                Heavy Metals ............................................................... 34
       2.3.2    Iron Oxide Nanoparticles as Adsorbent for
                Organic Contaminants ................................................. 37
       2.3.3    Iron Oxide Nanoparticles as Photo-Catalyst ............. 37
2.4    Challenges and Future Perspective of Using Iron Oxide
       Nanoparticles ................................................................................ 38
2.5    Conclusion .................................................................................... 41
Acknowledgments ................................................................................... 42
References ................................................................................................ 42

**Chapter 3**  TiO$_2$ Doped Lignocellulosic Biopolymer for Water Treatment:
              Waste to Wealth – A Pathway Towards Circular Economy ............... 47

*Abu Nasser Faisal, Zaira Zaman Chowdhury, Masud Rana,
Ahmed Elsayid Ali, Rahman Faizur Rafique and Rafie Bin Johan*

3.1    Introduction .................................................................................. 47
3.2    Synthesis of TiO$_2$/Cellulose and its Derivatives Based
       Composites for Wastewater Purification ...................................... 51
       3.2.1    TiO$_2$ Coating over the Cellulosic Filaments ............... 51
       3.2.2    Cellulose Derived Carbon as Support Matrix
                of TiO$_2$ ........................................................................ 52
       3.2.3    Porous TiO$_2$ Based Composites Using Cellulosic
                Filament as Sacrificial Templates ............................... 52
3.3    Synthesis of TiO$_2$/Carbon and its Derivatives Based
       Composites for Wastewater Purification ...................................... 54
3.4    Photocatalytic Degradation Performance by TiO$_2$ Doped
       Cellulosic Matrix for Water Treatment ........................................ 55
3.5    Photocatalytic Degradation Performance by TiO$_2$ Doped
       Carbonaceous Substrate for Water Treatment .............................. 56

3.6     Recent Roadmap, Challenges and Future Perspective of
        TiO$_2$ Doped Biopolymer Composites ..........................................58
3.7     Conclusion .........................................................................58
Acknowledgments .........................................................................60
References .........................................................................60

**Chapter 4**    Electrospun Nanofibre/Composite Membrane
                 for Water Treatment.........................................................63

*Bagavathi Krishnan, Zaira Zaman Chowdhury, Masud
Rana, Ahmed Elsayid Ali, Rahman Faizur Rafique, Amutha
Chinnapan and Md. Mahfujur Rahaman*

4.1     Introduction .........................................................................63
4.2     Basic Principle for Electrospun Membrane Fabrication
        for Water Treatment.............................................................65
4.3     Types of Electrospun Membrane Fabrication Process ...........68
        4.3.1    Fabrication of Tri-Axial Electrospun Fibres for
                 Water Filtration Membrane ......................................68
        4.3.2    Fabrication of Co-Axial Electrospun Fibres for
                 Water Filtration Membrane ......................................69
        4.3.3    Fabrication of Electrospun Fibres for
                 Water Filtration Membrane by Coupling of
                 Electrospinning with Electro-Spraying....................70
        4.3.4    Fabrication of Electrospun Fibres for Water
                 Filtration Membrane Using Melt Phase Splitting
                 and Dispersion.........................................................70
4.4     Application of Electrospun Nanofibre and Its Composites
        for Pressure Driven Membrane Filtration Unit for Water
        Treatment.............................................................................70
        4.4.1    Reverse Osmosis (RO)..............................................71
        4.4.2    Nanofiltration (NF) .................................................71
        4.4.3    Ultrafiltration (UF)...................................................71
        4.4.4    Microfiltration (MF).................................................72
4.5     The Drawbacks of Utilising Electrospun Nanofibre and
        its Composites Wastewater Filtration/Treatment .................72
4.6     Conclusion .........................................................................74
Acknowledgments .........................................................................75
References .........................................................................75

**Chapter 5**    Green Nanotechnology in Wastewater Treatment............................79

*G. K. Prashanth, M. Mutthuraju, Manoj Gadewar, Srilatha
Rao, K. V. Yatish, Mithun Kumar Ghosh, A. S. Sowmyashree
and K. Shwetha*

5.1     Introduction .........................................................................79
5.2     Nanomaterials in Green Context.............................................80

5.3    Why Plant Extracts Are More Promising Than
       Organisms...................................................................................80
5.4    Types of Nanomaterials Used in Wastewater Treatment.........80
       5.4.1    Metal-Based Nanomaterials.....................................81
       5.4.2    Nanocatalyst.............................................................83
       5.4.3    Magnetic NPs ..........................................................83
       5.4.4    Nano Membranes .....................................................83
5.5    Wastewater Treatment by Green NPs.....................................84
       5.5.1    Filtration Technique .................................................84
       5.5.2    Pollutant Adsorption by NPs....................................84
       5.5.3    Organic Compounds Removal by NPs.......................85
       5.5.4    Heavy Metal and Ions Removal by NPs....................86
5.6    Future Plan ...........................................................................86
5.7    Summary ...............................................................................87
References .....................................................................................88

**Chapter 6**    Integration of Microbial Treatment for Advanced Biological
Treatment of Wastewater..................................................................95

*R. Reevenishaa Ravi Chandran, Zaira Zaman Chowdhury,
Masud Rana, Ahmed Elsayid Ali, Rahman Faizur Rafique,
Md. Mahfujur Rahaman and Karthickeyan Viswanathan*

6.1    Introduction .........................................................................95
6.2    Implications of Treating Wastewater in Present Scenario.......99
6.3    Multiple Streams of Wastewater from the
       Industrial Sector ..................................................................99
       6.3.1     Battery Industry ............................................................99
       6.3.2     Electrical Power Plant ................................................100
       6.3.3     Nuclear Power Plant ...................................................100
       6.3.4     Textile Industry ..........................................................101
       6.3.5     Pulp and Paper Industry ..............................................101
       6.3.6     Leather Industry ........................................................101
       6.3.7     Agricultural Industry...................................................101
       6.3.8     Pharmaceutical Industry .............................................102
       6.3.9     Oil Mill Industry .......................................................102
       6.3.10    Organic Chemical Production Industries.................102
       6.3.11    Petroleum and Petro-Chemical Industries ..............102
       6.3.12    Mines Industry ..........................................................102
       6.3.13    Dairy Industry ...........................................................103
       6.3.14    Steel and Iron Industry...............................................103
       6.3.15    Food Industry .............................................................103
6.4    The Hybrid Mechanism of Microalgal/Bacterial
       Approach towards the Purification of Wastewater ...............104
6.5    The Advantages and Disadvantages of the Hybrid
       Approach Using Microalgal/Bacterial Treatment for
       Wastewater Purification.......................................................105
6.6    Conclusion .........................................................................110

Acknowledgments .......................................................................... 112
References .................................................................................... 112

**Chapter 7**   A Comprehensive Analysis on Bio-Reactor Design and
Assessment Biological Pre-Treatment of Industrial
Wastewater ................................................................................ 117

*Santhana Sellamuthu, Zaira Zaman Chowdhury,
Masud Rana, Ahmed Elsayid Ali, Rahman Faizur Rafique
and Seeram Ramakrishna*

7.1    Introduction ................................................................. 117
7.2    History for Development of MBR Technology .................... 120
7.3    Types of Configuration/Design in MBR .............................. 121
7.4    Types of Membrane Bioreactor ......................................... 121
       7.4.1   MBR with Aerobic Stirred Reactor.......................... 121
       7.4.2   MBR with Anaerobic Stirred Reactor...................... 123
       7.4.3   MBR with Fluidized Bed Reactor............................ 124
       7.4.4   MBR with Semi-Fluidized Bed Reactor .................. 125
       7.4.5   MBR with DSFF Reactor......................................... 125
       7.4.6   MBR with Inverse Fluidization Bed Reactor.......... 127
7.5    Conclusion ................................................................... 128
Acknowledgments ..................................................................... 129
References .................................................................................... 129

**Chapter 8**   De-Ballasting Water Treatment Technology to Protect the
Sea-Ocean Environment from Ships at the Western
Australian Ports.......................................................................... 132

*Tien Anh Tran*

8.1    Introduction ................................................................. 132
8.2    Literature Review ......................................................... 134
       8.2.1   BWM ................................................................... 134
       8.2.2   The Ballast Water Treatment Technologies ............ 134
       8.2.3   Recent Works Related to the Ballast Water
               Treatment............................................................. 135
       8.2.4   Wastewater Treatment Technology ........................ 135
8.3    Case Study: Western Australia Ports................................. 137
8.4    Concluding Remarks ..................................................... 141
Acknowledgments ..................................................................... 141
References .................................................................................... 141

**Chapter 9**   Phytoremediation of Hg and Cd Particulates Using
Scenedesmus and Nostoc on Industrial Wastes ............................. 144

*Dr. M. Mathiyazhagan, Dr. G.Bupesh, Dr. P. Visvanathan
and P. Sudharsan*

9.1    Introduction ................................................................. 144

            9.1.1   Macrophytes .................................................. 144
            9.1.2   Scenedesmus ............................................... 145
            9.1.3   Nostoc ...................................................... 145
     9.2    Toxicity of Heavy Metals ...................................... 146
            9.2.1   Mercury .................................................... 147
            9.2.2   Cadmium .................................................. 149
     9.3    Conclusion ........................................................ 149
     References ................................................................. 150


**Chapter 10** Employment of Phytal Mediated Metallic Nanoparticles
            for the Effective Remediation of Wastewater Particulates .............. 153

     *P. Sudharsan, T. Siva Vijayakumar, G. Bupesh, Bhagyudoy*
     *Gogoi and M. Mathiyazhagan*

     10.1   Introduction ....................................................... 153
            10.1.1  Nanoparticle Synthesis and Wastewater
                    Treatment ................................................. 153
     10.2   Green Synthesis of Nanoparticles ........................ 156
     10.3   Conclusion ........................................................ 158
     References ................................................................. 158

**Index** ................................................................................. 161

# Preface

This book presents the green synthesis towards the treatment of wastewater from various sources and wastewater treatment using green technology that have a high potential for large-scale application in the near future. It explores utilising the novel biotechnological tools for achieving paths in bioremediation through microorganisms along with discussing the state-of-the-art development in treatment processes of toxic pollutants by application of bioremediation along with present and future perspectives. Additionally, this book represents the green nanotechnology which is used for improving water and wastewater treatment efficiency and augmenting water supply through the safe use of unconventional water sources. The use of green nanoparticles in wastewater treatment is a cost-effective, convenient, and environmentally beneficial option and reviews the fabrication of iron-based nanoparticles, including various polymorphs of iron oxides, oxyhydroxides, iron hydroxide, and zero-valent iron nanoparticles. Consequently, this book deals with the overview of previous advances in the synthesis of carbon-based photocatalysts derived from biomass, $TiO_2$ Doped Lignocellulosic Biopolymer with an emphasis on their photodegradation potential for a variety of pollutants, and evaluating the advantages related to the application of wastewater treatment utilising Electrospun Nanofibre/Composite Membrane. It also highlights the advanced biological treatment systems which approaches linked with permanent film growth which have been demonstrated to be significantly more effective and efficient than suspending growth systems and explores remediation of industrial sludges/effluents with a particular emphasis on biological treatment, including the construction of bioreactors and addressing the use of membrane-based and liquid phase oxygen-based technologies in wastewater treatment, even aerobic and anaerobic procedures. The application of de-ballasting water treatment technology is to protect the sea-ocean environment from ships by evaluation and pre-treatment of the ballast water in order to decrease and eliminate the harmful species from discharging procedures. Furthermore, this book represents the phytoremediation of toxic heavy metals from industrial wastewater using microbes like scenedesmus and nostoc with the ability to utilize nutrients and reduce the risk of eutrophication and represent the green nanoparticles for the treatment of the wastewater present in the environment. Finally, this book gives out new trends and advances in wastewater treatment in different aspects with detailed discussions on recent advancements of various technologies. Each chapter highlights the scope and prospects of the technically advanced treatment in achieving the sustainable development in multi-disciplines. This book highlights these techniques in detail and presents the state-of-the-art in this field and the opportunities these techniques offer to improve process performance.

**Swapnila Roy**
**Tien Anh Tran**
**Zaira Zaman Chowdhury**
**Prathibha B.S.**

# About the Editors

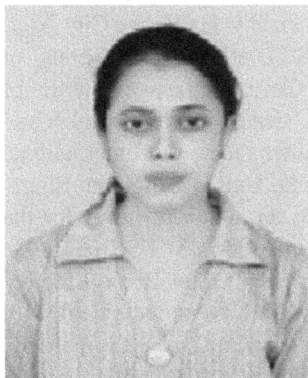

**Dr. Swapnila Roy** is Associate Professor and HOD of School of Basic and Applied Sciences in Lingaya's Vidyapeeth, Faridabad, Haryana. Before that she worked as Research Director in OPJS University, Churu, Rajasthan. Prior to that she worked as Associate Professor (Chemistry) in the department of Natural and Applied Sciences, Glocal University, Saharanpur, UP. Before that she worked as Assistant Professor in the Dept. of Chemistry (School of Applied Sciences), K K University, Nalanda, Bihar. She has been awarded a PhD from Jadavpur University, Kolkata in 2018. Prior to KKU, she worked at Regent Institute of Science and Technology, Kolkata, India as a lecturer. She worked as Scientific Assistant in the World Bank Funded Project (7 years) in West Bengal Pollution Control Board. She worked as Editor in a reputed publishing house (Kolkata). She also worked as Analytical Chemist in different pharmaceutical and chemical companies. She received her PhD in Environmental Chemistry from Jadavpur University, Kolkata, India. Her primary areas of research mainly focused on defluoridation in wastewater, adsorption methods for remediation, and designing expert tools. Her research areas span over different pollutants separation and process intensification. She is skilled in handling sophisticated instruments such as ICP-OES, AAS, FT-IR, HPLC, GCMS, bomb calorimeter, Soxlet apparatus, and different instruments used for analysis of hazardous waste. She has authored 34 reputed journals and book chapters and has been published in international conferences. She has 2 patents to date and some patents are ready to be filed soon. She received a Young Research Excellence Award, an International Scientist Award, and Researcher Award for outstanding performances. She has edited a few books with other publications. She is a reviewer of some reputed journals such as the *American Journal of Chemical Research, International Journal of Environmental Chemistry, CleanTechnologies and Environmental Policy, Groundwater for Sustainable Development*, and the *Global Research and Development Journal*. She has been invited as a speaker at some FDP and STTP programs in different engineering colleges.

**Dr. Tien Anh Tran** is Assistant Professor at Department of Marine Engineering, Vietnam Maritime University, Haiphong City, Vietnam. He graduated with a BEng. and MSc in marine engineering from Vietnam Maritime University, Haiphong City, Vietnam. He received his PhD at Wuhan University of Technology, Wuhan City, People's Republic of China in 2018. He has been invited as a speaker and an organisation committee member of international conferences in Australia, United Kingdom, Singapore, China, etc. He is an author/reviewer for the international journals indexed in SCI/SCIE, EI. He is Leading Guest Editor and Editor Board Member for various journals. In 2021, he edited a book with another publication house. His current research interests include ocean engineering, marine engineering, environmental engineering, machine learning, applied mathematics, fuzzy logic theory, multi criteria decision making (MCDM), maritime safety, risk management, simulation and optimisation, system control engineering, renewable energy and fuels, etc.

**Dr. Zaira Zaman Chowdhury** is working as Assistant Professor and Industrial Training Co-ordinator in the Nanotechnology and Catalysis Research Centre (NANOCAT), a potential National HICOE and UM COE. At present she has 12 PhD scholars and 11 MPhil scholars under her supervision. Dr. Zaira has received several esteemed awards such as the Excellent Scientific Award in 2020 for "Carbon Chemistry and Water Treatment"; the IMRF Brezileius Award in Nanotechnology and Catalysis 2019, India; Best Researcher Award in Applied Chemistry and Chemical Engineering 2019, India; ISPA Gunasekaran Award 2018, India; Best Scientist Award in Natechnology 2018, India; the Atlas award in 2015 awarded by another publishing house; Gold Award and Silver Award from Malaysian Inventions and Design Society (MINDS) and University Malaysia Perlis (UniMAP, 2013). She has received CIMA Appreciation Award from Malaysian Inventions and Design Society (MINDS) and University Malaysia Perlis (UniMAP, 2013). Recently she has received the Bronze Award from International Engineering Innovation and Invention Exhibition (IENVEX), University Malaysia Perlis (UniMAP 2014). In 2016, she received the best paper award on Fourth Generation Nano-antioxidant Synthesis from New Zealand. She has worked for synthesis of micro and nano structured carbon and biomass processing using catalyst. She has contributed more than 101 papers in *ISI* and *Scopus* journals, has 2 international patents with 85 international Conference Proceedings; and has an H index 23 and a Google scholar H index of 25.

Earlier she was appointed as Senior Research Fellow in the same institution. She completed her post-doctoral research fellowship from 1/10/2013–31/03/2015 at the Nanotechnology and Catalysis Research Centre (NANOCAT), University Malaya. Previously she was appointed as Research Officer at the Nanotechnology and Catalysis Research Centre (NANOCAT), University Malaya from 1/04/2013 to 30/09/2013. She started her career in University Malaya from November 2008 as Research Assistant in the department of chemistry. She received her PhD in environmental analytical chemistry and engineering from University of Malaya in 2013 and her MSc. in polymer engineering and technology from the University of Dhaka, Bangladesh in 2003 and her BSc in applied chemistry and chemical engineering technology from University of Dhaka, Bangladesh in 2001 and her bachelor of education (BEd) in 2007 from Dhaka, Bangladesh. She received the Cambridge International Diploma for Teacher &and Trainers (CIDTT) from Cambridge University, UK with distinction in 2006–2007. She has experience in development of fibre polymer composites, wastewater treatment, and adsorbent preparation for batch and fixed bed adsorption systems.

**Dr. Prathibha B.S.** obtained her MSc from Central College Bangalore and her PhD from Kalasalingam University, Tamil Nadu, India. She is currently working as professor and head in the Department of Chemistry, BNM Institute of Technology, Bengaluru. She has 22 years of teaching and 10 years of research experience. She has published 23 research articles in peer-reviewed journals and has 5 international and 1 national technical patent to her credit. She has filed 6 patents. She has authored a book titled *Quaternary ammonium compounds as corrosion inhibitor.* Her area of research includes material science, corrosion, nanotechnology, and fuel cells. She was the recipient of the Indira Priyadarshini award in 2013, the "Education Leadership Award-2022" by Glantor X Market Research and Ranking, Best Researcher of the Year Award 2020 by INSc, and Best Chemistry Professor of the year 2020 at the International Education Award Conference in 2020, by Kites Education. She also received the Global Digital Teachers Excellence Award in 2020 from the Global Digital Academy, a unit of CK International Business Solutions PVT LTD.

# Contributors

**Vidya Acharya**
Department of Environmental
  Science
University of Calcutta, Kolkata,
  WB, India

**Ahmed Elsayid Ali**
Nanotechnology and Catalysis
  Research Center (NANOCAT)
University of Malaya
Malaysia

**G. Bupesh**
Department of Forestry
Nagaland University
Nagaland

**R. Reevenishaa Ravi Chandran**
Nanotechnology and Catalysis
  Research Center (NANOCAT)
University of Malaya
Malaysia

**Zaira Zaman Chowdhury**
Nanotechnology and Catalysis
  Research Center (NANOCAT)
University of Malaya
Malaysia

**Amutha Chinnapan**
Center of Nanotechnology and
  Sustainability
Singapore National University
Singapore

**Abu Nasser Faisal**
Nanotechnology and Catalysis
  Research Center (NANOCAT)
University of Malaya
Malaysia

**Manoj Gadewar**
School of Medical and Allied Sciences
KR Mangalam University
Gurgaon

**Mithun Kumar Ghosh**
Department of Environmental Science
Government College Hatta
Damoh

**Bhagyudoy Gogoi**
Department of Environmental Science
Nagaland University
Nagaland

**Alli Malar Harikrishnan**
Nanotechnology and Catalysis Research
  Center (NANOCAT)
University of Malaya
Malaysia

**Venkatalakshmi Jakka**
Department of Chemistry, School of
  Applied Sciences and Humanities,
  Vignan's Foundation for Science,
  Technology and Research (VFSTR,
  deemed to be University)
Vadlamudi, Guntur, AP, India

**Rafie Bin Johan**
Nanotechnology and Catalysis Research
  Center (NANOCAT)
University of Malaya
Malaysia

**Bagavathi Krishnan**
Nanotechnology and Catalysis Research
  Center (NANOCAT)
University of Malaya
Malaysia

**M. Mathiyazhagan**
Department of Botany
Silapathar Science College
Silapathar, Dhemaji, Assam-787059.

**Ajita Mitra**
Department of Public Health, University
    of D Nanotechnology and Catalysis
    Research Center (NANOCAT),
    University of Malaya
Malaysia Haka, Bangladesh

**Sai Rupa Mopidevi**
Department of Chemistry, School of
    Applied Sciences and Humanities,
    Vignan's Foundation for Science,
    Technology and Research (VFSTR,
    deemed to be University)
Vadlamudi, Guntur, AP, India

**Aniruddha Mukhopadhyay**
Department of Environmental Science
University of Calcutta
Kolkata, WB, India

**M. Mutthuraju**
Department of Chemistry
Sai Vidya Institute of Technology
Bengaluru

**G. K. Prashanth**
Department of Chemistry
Sir M. Visvesvaraya Institute of
    Technology
Bengaluru

**Rahman Faizur Rafique**
Research and Environmental Consultant
RR Environmental, Malta
New York, USA

**Md. Mahfujur Rahaman**
Department of Commerce
Islamic Business School, Universiti
    Utara Malaysia (UUM)
Malaysia

**Srilatha Rao**
Department of Chemistry
Nitte Meenakshi Institute of Technology
Bengaluru

**Seeram Ramakrishna**
Center of Nanotechnology and
    Sustainability
Singapore National University
Singapore

**Masud Rana**
Nanotechnology and Catalysis Research
    Center (NANOCAT)
University of Malaya
Malaysia

**Arthita Ray**
Department of Environmental Science
University of Calcutta
Kolkata, WB, India

**Santhana Sellamuthu**
Nanotechnology and Catalysis Research
    Center (NANOCAT)
University of Malaya
Malaysia

**Shubhalakshmi Sengupta**
Department of Chemistry
School of Applied Sciences and
    Humanities, Vignan's Foundation for
    Science, Technology and Research
    (VFSTR, deemed to be University)
Vadlamudi, Guntur, AP, India

**K. Shwetha**
Department of Chemistry
Nitte Meenakshi Institute of Technology
Bengaluru

**A. S. Sowmyashree**
Department of Chemistry
Nitte Meenakshi Institute of Technology
Bengaluru

**P. Sudharsan**
Department of Forestry
Nagaland University
Nagaland

**Anjali Syamala**
Department of Chemistry,
  School of Applied Sciences and
  Humanities, Vignan's Foundation for
  Science, Technology and Research
  (VFSTR, deemed to be University)
Vadlamudi, Guntur, AP, India

**Tien Anh Tran**
Department of Naval Architecture and
  Ocean Engineering
Seoul National University, Seoul
South Korea
School of Computing Science and
  Engineering
Galgotias University
Greater Noida, India

**T. Siva Vijayakumar**
PG and Research Department of
  Biotechnology
Srimad Andavan Arts and Science
  College (A)
Tiruchirappalli, Tamil Nadu, India.

**P. Visvanathan**
Department of Zoology
ISBM University, Nawapara (Kosmi)
Block and Tehsil- Chhura, Gariyaband,
  Chhattisgarh

**Karthickeyan Viswanathan**
Department of Chemical Engineering
National Cheng Kung University
Taiwan

**K. V. Yatish**
Centre for Nano and Material Sciences
Jain University
Bengaluru

# 1 Bioremediation of Toxic Pollutants

*Vidya Acharya, Venkatalakshmi Jakka,*
*Arthita Ray, Sai Rupa Mopidevi, Anjali Syamala,*
*Aniruddha Mukhopadhyay and*
*Shubhalakshmi Sengupta*

## 1.1 INTRODUCTION

In the past few decades to meet the demand of the rising global population, there has been over-utilisation of natural resources. These exploitations exposed the environment to many contaminants. A few such incidents of contamination, for example, the Bhopal gas tragedy, Fukushima incident, Chernobyl incident, etc., have polluted the surroundings beyond repair. Moreover, other anthropogenic and industrial activities also generate contaminants such as heavy metals, dyes, and other organic compounds which pose a great threat to the ecosystem. Dyes which are considered hazardous by the Environmental Protection Agency along with heavy metals are released into the water bodies as effluent from various industries, especially the textile industry [1, 2]. Studies show that approximately worldwide, 100 tons of dyes are released into the water bodies per year. [3, 4]. These contaminants are extremely harmful due to their carcinogenic properties. Thus its removal is a necessary step to restore the ecosystem. Apart from dyes and heavy metals various organic pollutants, pharmaceutical wastes, microplastics, and radioactive wastes are toxic to the living organisms. Control of pollution and removal of these toxicants as a consequence have become the need of the hour. Remediation of these toxicants through various physical and chemical processes are often not environmentally sustainable and development of those toxicant removing materials and processes always do not follow a green route. But these processes if done biologically using microorganisms can impart effective environmental sustainability to the process. Thus, bioremediation of such toxicants has been studied by various researchers worldwide. These works have been discussed in this chapter.

### 1.1.1 BIOREMEDIATION

The term bioremediation has two parts: *bios* meaning life and *remediate* implying solving. In other words, bioremediation is the technique in which different kinds of biological organisms are used to degrade the contaminants that pollute the environment.

It is a newly evolved technology which is used on a large scale to detoxify contaminants from the environment. Depending upon which type of organism is used

DOI: 10.1201/9781003342830-1

for bioremediation, it can be subcategorised into phytoremediation (by the use of plants), mycoremediation (by the use of fungi), and microbial remediation (by the use of bacteria).

### 1.1.1.1  Types of Bioremediation

Bioremediation can occur on-site and off-site. Depending upon this it can be classified into in situ bioremediation and ex situ bioremediation.

#### 1.1.1.1.1  In Situ Techniques

These techniques are used on site and treat the contaminants without causing any major disturbance to the original ecosystem [5]. Much literature suggests that this technique has treated dyes, heavy metals, and other solvents, yet there are some parameters which are importantly maintained in order to conduct an in situ bioremediation. Some of these parameters are pH, temperature, humidity, nutrient availability, etc. [6] Moreover, in the case of in situ bioremediation of soil, porosity is the most important factor. [5]

- *Bioaugmentation*: Microbial cultures are added to the site which is to be remediated. This procedure is the most common among in situ bioremediation.
- *Bioventing*: This method uses controlled aeration of the vadose zone so as to increase the process of bioremediation by increasing the activities of microbes.
- *Bioslurping*: It uses a combination of bioventing, aeration of groundwater along with pumping a using vacuum pump. It is used to bio-remediate even the volatile substances.
- *Biosparging*: In this technique, air is injected to the saturated zone and the whole process is very much similar to the bioventing technique.
- *Biofilters*: This technique helps in remediating toxins in the air columns.

#### 1.1.1.1.2  Ex Situ Techniques

In the ex situ bioremediation techniques, the pollutants are transported from the site to the treatment location. Important parameters for these types of technique are geographical morphology of the site, type of pollutant, intensity of pollution, and cost of remediation [6].

- *Bioreactors*: A bioreactor is a container which uses a set of reactions and in which the remediation takes place. The conditions inside the bioreactor are checked so as to provide optimum environment for the microbial culture. The main disadvantage of this method is, when used on a large scale, a bigger set of bioreactors is needed which may not always be a cost-effective option.
- *Land farming*: It is one of the most economical ways of bioremediation which includes transport and tillage so as to increase aeration. This method requires a large area for operation and cannot be used if volatile substances are present.

- **Biopile**: In this method, the contaminated soil is transported and the nutrient is added and aerated in order to increase the microbial activity. This is an effective method in the low temperature region. [7] Being a flexible process, the remediation time can be controlled by the use of a heating system but excess use of heating can lead to dryness, which may inhibit the microbial activities, thus reducing the bioremediation effectiveness. [8]
- **Windrows**: In this method, the polluted soil is turned over along with addition of water, periodically with the intention of increasing aeration levels and enhances the microbial degradation. This method, however, cannot be used in case of volatile pollutants. [9]

## 1.2 DYES

Dyes are complex coloured organic compounds and are extensively used in textile, cosmetics, printing industries, etc. The property of the dye to stain the medium is due to the interaction of its two components – auxochrome and chromophore. [10] Auxochrome is an electron donating group; meanwhile chromophore is an electron accepting group and when these two interplay, the conjugated system thus formed retains and stains the media. Dye can be produced naturally and also from synthetic sources. [10]

Natural dyes are made from natural sources, for example, orange colour from saffron and blue colour from indigo plant, along with insects and minerals to produce colour without any additional chemical treatment. [11] Being of natural origin, these dyes are costly to manufacture and in addition have a poor colour fastness property so they require mordant which are generally heavy metals used in the industries. [12]

Synthetic dyes are both a boon and a bane to the human society. Synthetic dyes colour the textile world for humans, improving and providing for their needs, but on the other hand, the non-biodegradable compounds of the dyes cause pollution to the environment. Moreover the majority of these compounds are carcinogenic and cause allergic reactions in humans. The first synthetic dye, picric acid, was synthesised by Woulfe in the year 1871 while accidently treating indigo dye with nitric acid. [13]

### 1.2.1 CLASSIFICATION OF SYNTHETIC DYES

Classification of dyes is mostly based on the chemical structure and composition. Among all the types of synthetic dyes the most common is azo dyes having $-N = N$-group, and they are used in a variety of industries including, textiles, cosmetics, food, paper, etc. It is summarised in Table 1.1

## 1.3 BIOREMEDIATION OF DYES

Bioremediation of dyes includes use of various enzymes and microbes such as algae, bacteria, yeast, and fungi. These microbes disable the movement of chemical structure of synthetic dyes eventually leading to degradation and mineralisation processes. [14, 15]

**TABLE 1.1**
**Summary of Different Types of Synthetic Dyes**

| Types of Dyes | Examples of Dyes | Application |
|---|---|---|
| Cationic dyes/acid dyes (containing metal complexes) | Congo red, methyl orange acid and triphenylmethane | Wool, nylon, silk |
| Anionic dyes/basic dyes | Basic red, methylene blue, crystal violet, aniline yellow, brilliant green, hemicyanine, cyanine, diazahemi cyanine, azinediphenylmethane, xanthene, triarylmethane, and anthraquinone | Polyester, wool, silk, nylon |
| Sulphur dyes | Sulphur black | Cellulosic fibres |
| Azoic dyes | Mono azo, diazo, and triazo dyes | Most of the textiles, paper, cosmetic, and food industry |
| Direct dyes | Martius yellow, direct black, orange, blue, violet, red, phthalocyanine, oxazine | Cotton, wool, silk |
| Dispersive dyes | Disperse blue, disperse red disperse orange, disperse yellow, disperse brown, benzodifuranone, and styryl | Nylon, polyester |
| Vat dyes | Benzanthrone, vat blue, vat green, indigoids, and anthraquinone | Wool, rayon fibres |
| Reactive dyes | Reactive (red, blue, yellow, black), remazol (blue, yellow, red, etc.) formazan and phthalocyanine | Cellulosic fibres, wool |

### 1.3.1 BY THE USE OF BACTERIA

Studies have concluded a total degradation of synthetic dyes by the use of bacteria. [16] Most of the studies found in literature have a consistent set of parameters of biodegradation such as a particular temperature, pH, and concentration of dyes and time of experiment. A brief summary of the various studies conducted in recent years is presented in Table 1.2.

### 1.3.2 BY THE USE OF FUNGI

Different fungal species are used for the remediation and processes involved are azo bond reduction, desulphonation, deamination, etc. The enzymes present in the fungal species adsorb the dye followed by subsequent degradation. [25] There are a few factors that affects the degradation process of dyes including pH, temperature, concentration of nutrient in the media, oxygen saturation level, type of dye, and growth rate of the fungal species [26]. The Table 1.3 summarises the existing literature on the bioremediation of dyes by the use of various fungal species.

**TABLE 1.2**
**Summary of Studies on Bioremediation of Dyes Using Bacteria Species**

| Bacteria species | Dye | Degraded products | Conditions | | | | Mechanism | Reference |
|---|---|---|---|---|---|---|---|---|
| | | | Temperature (°C) | pH | Time (hours) | Conc. of dyes (mg/L) | | |
| Bacillus aryabhattai | Coomassie Brilliant blue G-250 (CBB), Indigo carmine, and remazol Brilliant blue R (RBBR) | | 37 | 5–8 | 72 | 150 | Azo reduction | [16] |
| Bacterial consortium of Pantoea Ananatis, Bacillus Fortis, Alcaligenes faecalis, Brevibacillus parabrevis and Bordetella trematum | Acid Blue (193 and 194) | 4-amino-3-hydroxynaphthalene-1- sulfonic acid, 2-aminonaphthalene-1-ol, naphthalene-1,2-diol, pyruvic acid, acetaldehyde, 3-hydroxy-7-nitro-naphthalene-1-sulfonic acid | 35 | 7.4 | 96 | 100 | Bioreduction | [17] |
| Mesophilic and thermophilic lactic acid bacteria (LAB) | Dorasyn red azo dye | | 20 | 3 | 3 | 100 | Biosorption | [18] |
| Pseudomonas extremorientalis | Cresol red | | 30 | 8 | 24 | 50 | Redox reaction | [19] |
| Streptomyces ipomoeae | Cresol red | | 35 | 8 | 24 | 50 micrograms | Redox reaction | [20] |
| Bacillus megaterium, Micrococcus luteus and Bacillus pumilus | Reactive black | Anthraquinone, hydroxy and amino derivative | 30 | 7 | 120 | 25 | Azo reduction | [21] |
| Aeromonas hydrophila | Reactive black 5 | Aromatic amines | 35 | 7 | 24 | 100 | Cleavaging and reduction of aromatic amines | [22] |
| Halomonas sp. | Reactive red 184 (RR184) | | | 8.5–11.5 | | | Dye mineralisation | [23] |
| DDMY1 | Reactive black 5 | | 30–40 | 4–9 | 48 | | Decolourisation process | [24] |

**TABLE 1.3**

**Summary of Literature on Bioremediation of Dyes Using Fungal Species**

| Type of Fungi | Dye | Degraded Products | Conditions | | | | Mechanism | Reference |
|---|---|---|---|---|---|---|---|---|
| | | | Temperature (°C) | pH | Time (hours) | Conc. of dyes (mg/L) | | |
| Laccase from *Peroneutypa scoparia* | Acid red 97 | Naphthalene 1,2-dione and 3-(2-hydroxy-1-naphthylazo) benzenesulfonic acid | 40 | 6 | 6 | 100 | Reduction of azo bond | [27] |
| *Aspergillus niger* | Red azo dye | | 28 | 9 | 2 | 500 | Breaking the dye complex bonds | [28] |
| *Lasiodiplodia* sp. | Malachite green | | | | | | Oxidation and mineralisation | [29] |
| *Aspergillus niger* | Acid orange 56, acid blue 40 and methyl blue | | | | | | Biosorption | [30] |

### 1.3.3 BY THE USE OF ALGAE

Various studies have been conducted to degrade the complex organic and inorganic compounds with the help of different algal species. A brief summary of literature is provided in the Table 1.4.

### 1.3.4 BY THE USE OF YEAST

The table 1.5. summarises the existing literature on the bioremediation of dyes by the use of yeast.

### 1.3.5 BY THE USE OF ENZYMES

Enzymes or biocatalysts have recently become very popular among various industrial activities. These are used in the remedial process to remove the toxic substances by precipitation method. [35] The parameters needed for the remediation are pH, temperature, concentration of dyes and enzymes, type of dyes, etc. [36] Immobilisation of enzymes for the bioremediation is a crucial step and this is successfully achieved by cross-linkage. [37] Depending upon the type of dye and enzyme, various techniques are used for the remediation. Enzymes are known to degrade most of the known azo and non-azo dyes. In a recent study, it was proved that horseradish peroxidase successfully removed on an average 72% of different types of synthetic dyes [38] while another study used laccase to degrade 82% of Indigo Carmine dye. [39] Table 1.6 summarises the existing literature on the bioremediation of dyes by the use of various enzymes.

## 1.4 HEAVY METALS

Heavy metals pollution is a critical problem to the environment because of its presence in wastewater, groundwater, and soil. Heavy metals are inorganic contaminants having a specific density of 5 g cm$^{-1}$ and are not easily degradable. [42] The major source of heavy metals comes from anthropogenic activities such as industrialisation and urbanisation. [43] Some of these metals such as arsenic (As), cadmium (Cd), lead (Pb), and mercury (Hg) enter the human body through oral and respiratory pathways and are lethal in very low concentration. [44] When ingested in high concentration, heavy metals either interfere with normal metabolic processes by forming stable metal-organic complexes such as phytochelatins and metallothioneins [45] or develop resistance to these metals. [46] Some striking case studies include Itai-Itai disease (due to cadmium), the Minamata disease (due to mercury) of Japan, the black foot disease (due to arsenic) of China, etc. which lead to widespread deaths.

Various prior studies have successfully conducted the degradation of these metals by different physio-chemical processes yet these procedures due to being costly could not be effectively implemented on a large scale. Hence, in recent times, its degradation using a diverse set of bio-remedial techniques is being explored and conducted. Several bacteria, fungi, algae, and plant species have the ability to take up these metals from their culturing media henceforth contributing towards remediation. [42]

**TABLE 1.4**
**Summary of Literature on Bioremediation of Dyes Using Algal Species**

| Type of Yeast | Dye | Degraded Products | Conditions | | | | Mechanism | Reference |
|---|---|---|---|---|---|---|---|---|
| | | | Temperature (°C) | pH | Time (hours) | Conc. of dyes (mg/L) | | |
| *Dermatocarpon vellereceum* | Direct black 22 | Alkanes, sulphides, charged amines, sulphoxides, substi-tuted benzene rings | 40 | 8 | 28 | 50 | Enzymatic degradation | [31] |
| *Scenedesmus* | Methylene blue | | | | | | Biosorption | [32] |

**TABLE 1.5**

**Summary of Literature on Bioremediation of Dyes Using Yeast**

| Type of Yeast | Dye | Degraded Products | Conditions | | | | Mechanism | Reference |
|---|---|---|---|---|---|---|---|---|
| | | | Temperature (°C) | pH | Time (hours) | Conc. of dyes (mg/L) | | |
| *Scheffersomyces spartinae* | Acid scarlet 3R | 4-aminonaphthalene-1-sulfonic acid, 7,8-dihy-droxynaphthalene-1,3-disulfonic acid, and naphthalene-1,2,6,8-tetraol | 30 | 6 | 12 | 80 | Desulfonation, reductive cleav-age of azo bond | [33] |
| *Trichosporon akiyoshidainum* | Reactive black 5 | Aromatic amines, 5-amino-4-hydroxy-3,6-di-oxo-2,3, 5,6-tetra-hydro-naphthalene-2,7-sodium disulfonate | 25 | 7 | 12 | 200 | Biodegradation | [34] |

**TABLE 1.6**
**Summary of Literature on Bioremediation of Dyes Using Enzymes**

| Type of enzymes | Dye | Degraded products | Conditions | | | | Mechanism | Reference |
|---|---|---|---|---|---|---|---|---|
| | | | Temperature (°C) | pH | Time (hours) | Conc. of Dyes (mg/L) | | |
| Manganese Perxidase from *Irpex lacteus* F17 | Malachite green | 4-Dimethylaminobenzaldehyde, 4-(dimethylamino) benzophenone | 40 | 3.5 | 1 | 200 | Reduction of azo bond | [40] |
| Citrus limon peroxidase | Direct yellow | Carboxylic acid | 50 | 5 | | 18.75 | Radical initiated ring | [41] |

## 1.5   HEAVY METAL BIOREMEDIATION

In the race for survival, microbes have developed bioaccumulation techniques for heavy metals. This property is being exploited and used in the bioremediation processes.

### 1.5.1   BY THE USE OF BACTERIA

Bacteria are ubiquitously present in the environment and can survive even the harshest conditions. These microbes help in remediation of heavy metals through processes such as bioaccumulation, biosorption, etc. An elaborated summary of the studies conducted on bioremediation of heavy metals using biosorption technique by the use of bacteria is presented in Table 1.7.

### 1.5.2   BY THE USE OF FUNGI

Fungal species helps in sequestration of heavy metals in biomass. Studies show that they can easily survive in heavy metal contaminated area by accumulating it within their bodies. Glycoprotein and polysaccharides present in their cell wall binds with these heavy metals [52]. A few such studies are briefly described in the Table 1.8.

### 1.5.3   BY THE USE OF ALGAE

Algae can adsorb heavy metals in a short span of time and this slowly enters the cytoplasm by absorption process thus helping in their removal from the culturing media. [59] *Chlorella vulgaris* is known to remediate heavy metals such as lead (Pb) and cadmium (Cd) [60]

## 1.6   ORGANIC POLLUTANTS

Environmental pollution increased by several human activities such as industrialisation, mining, natural gas production, crude oil extraction, etc. the main organic pollutants were aromatic and hetero aromatic chemicals like pyridine, phenol, dimethyl phthalate, P-nitrophenol were causing several environmental problems. [61] Several

**TABLE 1.7**
**Summary of Studies of Bioremediation of Heavy Metals Using Bacteria Species**

| Bacteria Species | Type of metal | Reference |
|---|---|---|
| *Bacillus* sp. | Arsenic | [47] |
| *Geobacter sulfurreducens* | Chromium | [48] |
| *Pseudomonas aeruginosa* | Cadmium | [49] |
| *Pseudomonas aeruginosa* | Arsenic | [50] |
| *Phodopseudomonas palustris* | Cobalt | [51] |

**TABLE 1.8**
**Summary of Studies of Bioremediation of Heavy Metals Using Fungi**

| Fungal species | Type of metal | Reference |
|---|---|---|
| *Aspergillus flavus* | Arsenic | [53] |
| *Agaricus bisporus* | Cadmium | [54] |
| *Alternaria alternate* | Cadmium | [55] |
| *Aspergillus flavus* | Cadmium | [56] |
| *Aspergillus sp.* | Cadmium | [57] |
| *Alternaria alternata* | Mercury | [55] |
| *Agaricus bisporus*<br>*Pleurotus oestreatus* | Lead | [58] |
| *Agaricus bisporus* | Lead | [54] |
| *Aspergillus flavus* | Lead | [56] |
| *Agaricus bisporus*<br>*Pleurotus oestreatus* | Chromium | [58] |
| *Agaricus bisporus* | Chromium | [54] |
| *Agaricus bisporus* | Cobalt | [54] |

techniques were developed to remove these harmful pollutants such as ozonation, solvent extraction, chlorination, membrane process, adsorption, coagulation, flocculation, and biological degradation. [62] Biological degradation is more favourable because this is environmentally friendly and cost-effective as well. The biotechnological technique called bioremediation which can involve by biologically mediated changes in pollutants and it leads to manage pollutants via mineralisation and detoxification. [63]

## 1.7 BIOREMEDIATION OF ORGANIC POLLUTANTS

The microbial bioremediation was more promising, effective, innovative, and also environmentally beneficial than the other bioremediation technologies. This technique was used to remove pollutants in natural process involved by aerobic and anaerobic microbes such as bacteria, fungi, and micro-algae.

### 1.7.1 BY THE USE OF BACTERIA

Microbes degrade pollutants through their own metabolic process for degradation and biotransformation. Hydrocarbons and pesticides were destroyed by aerobic bacteria enzymes named rhodococcus, pseudomonas, sphingomonas, alcaligenes, and mycobacterium. The bioremediation of Polychlorinated biphenyls, chloroform, and trichloroethylene were treated by anaerobic bacteria enzymes. [64] Hydrolase, proteases, laccases, lipases, and other genetically engineered microorganisms were used for the different classes of organic pollutants in the process of bioremediation. [65]

### 1.7.2 BY THE USE OF ALGAE

In the recent decades, the bioremediation of organic pollutants using micro-algae known as phycoremediation has become very interesting due to its high capability of biomass growth rate, $CO_2$ fixation, and capability to be used as feedstock for biofuel production. [66] The microalgae use organic substance as a primary source of carbon and nutrients for their growth and survival in the process of organic content biodegradation. Some of the microalgae including C. vulgaris, C. pyrenoidosa, O. tenuis, tertiolecta, Dunaliella and Closterium lunula, and Ochromonas Danica were using for the organic pollutant's bioremediation. [67]

## 1.8 MICROPLASTICS

Plastic being an essential constituent in most products, it is widely used in all over the world on a daily basis and as a result, continuous increase in plastic waste generation creates an alarming situation affecting both of terrestrial and marine environments as well as disrupting human health, wildlife and their habitat, and climatic condition. In this plastic pollution, a major contributor is microplastics, particles with a grain size of less than 5mm. [68] Based on the source of origin, microplastics (MPs) are of two types: primary MPs and secondary MPs; most microplastics are secondary in nature generated by degradation of bigger plastics like megaplastics (> 1m), macroplastics (< 1m), and mesoplastics (< 2.5cm). The potential sources of MPs from land are wear of tyres on the road, microfibres generated from textiles after washing, sludge from the wastewater treatment plant in the agriculture sector, cruise ships in tourism, and microbeads used in personal care products. [69] MPs are produced in sea by the activities of commercial shipping, tourist cruises generating textile fibres, and fisheries by wear and tear of nets, ropes, and other gear during use. [70] Adverse impacts of MPs on marine creatures are MPs accumulating in animal bodies by direct ingestion or absorption from the primary tropic level such as microplankton, zooplankton, endocrine disruption in fishes due to the exposure to leaching of chemical additives from MPs, alteration of animal behavior, and disruption in reproductivity. [71, 72] As microplastics are also present in the atmosphere, MPs are found in remote regions and the radiative effects of atmospheric MPs affecting climate through radiation absorbance and scattering have been detected. [73]

At this stage of MPs pollution, bioremediation plays a key role as a promising eco-friendly strategy to overcome the adversity. Microbial biofilm formed on the plastic surface has the ability to degrade plastics by excreting enzymes. Several bacteria, fungi, and alga have been identified to have the capacity of plastic degradation. The polymer types, physical and chemical properties, and abiotic factors like UV radiation, heat, and stress are important factors in the biodegradation of plastics. [74]

## 1.9 MICROPLASTICS BIOREMEDIATION

Microbial degradation of plastics is a widely used bioremediation method as it is economically viable and eco-friendly in nature. [75] The steps of biodegradation are: biodeterioration where abiotic factors like UV radiation, temperature, and secreted enzymes from microorganisms increase the polymer surface area; then in

**TABLE 1.9**
**Microalgal Based Bioremediation for the Removal of Organic Pollutants**

| S.no | Organic pollutants | Species | Reference |
|------|--------------------|---------|-----------|
| 1. | Phenanthrene | *Chlorella sorokiniana and Pseudomonas Migulae* | [77] |
| 2. | Bisphenol | *Monoraphidium braunii* | [67] |
| 3. | Prometryne | *Chlamydomonas reinhardtii* | [78] |
| 4. | Dimethomorph and pyrimethanil | *Scenedesmus quadricauda* | [79] |
| 5. | Lindane, naphthalene | *Chlamydomonas sp.* | [80] |
| 6. | Isoproturon | *Scenedesmus quadricauda* | [79] |

biofragmentation polymers gets degraded into monomers, and finally by the assimilation process metabolisation of monomers into $CO_2$, biomass and $H_2O$ by microbes is done. [76] This is summarised in Table 1.9.

### 1.9.1 BY THE USE OF BACTERIA

Several bacterial species from different ecological niches (*Pseudomonas sp, Bacillus sp,. Vibrio sp., Ideonella sp., Staphylococcus sp.*) having plastic degrading ability have been identified. Instead of using a pure bacterial culture, researchers are using microbial communities consisting of diverse bacteria having distinctive physiological and metabolic capacities to degrade plastics. [81] A few studies on plastic biodegradation by bacteria are summarised in Table 1.10.

### 1.9.2 BY THE USE OF FUNGI

Adaptability in changing environments and the digestion capacity of environmental pollutants make filamentous fungi a key contributor in plastic degradation. Through aerobic or anaerobic digestion, they convert adsorbent nutrients (plastic polymers) to microbial biomass, $CO_2$, $H_2O$, and $CH_4/H_2S$. Various genera of fungi have been identified with the capability of plastic degradation. [81] A few studies on plastic biodegradation by fungi are summarised in Table 1.11.

### 1.9.3 BY THE USE OF ALGAE

Algae takes part in biofilm formation on plastic surfaces by secreting extracellular mucilaginous polymeric substances able to adhere and colonise. *Anabaena spiroides* (blue-green alga), *Scenedesmus dimorphus* (green microalga), and *Navicula pupula* (diatom) species were identified as having capacity to degrade polyethylene. [98] A diatom species, *Phaeodactylum tricornutum*, was used as a solution to decompose PET waste by secreting a large amount of PETase. [99] The functional expression of PETase in green algae *Chlamydomonas reinhardtii* was demonstrated in a study by co-incubation of algae cells able to express PETase with PET samples for 4 weeks at 30°C where a completely degraded form (Terephthalic acid) was found after incubation. [100] The current strategies of plastic bioremediation by photosynthetic

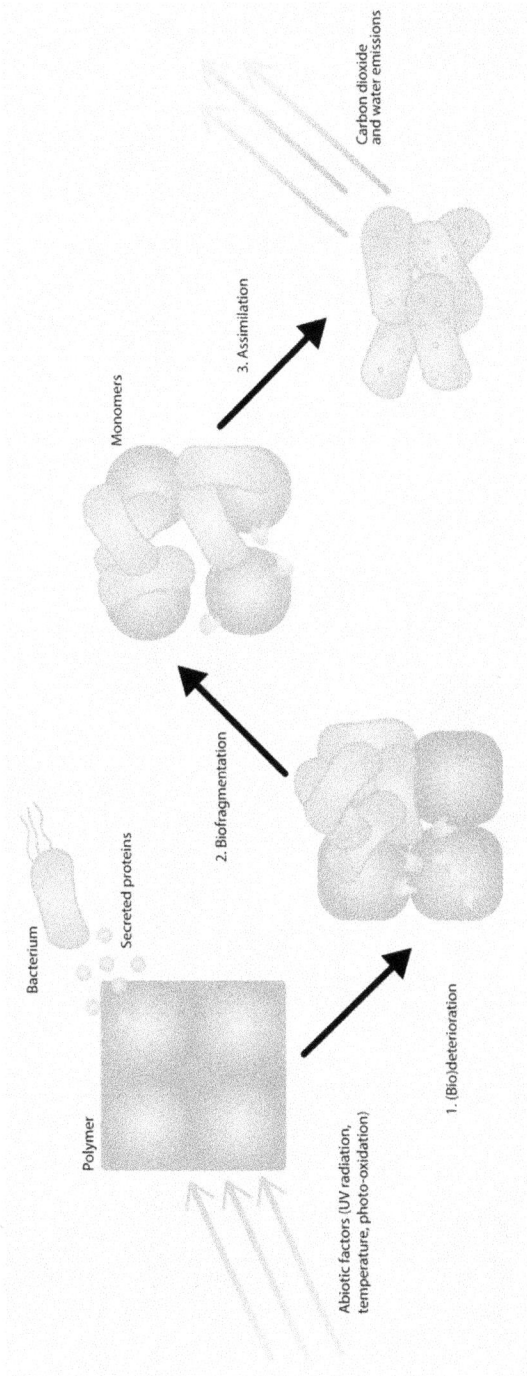

**FIGURE 1.1** Microbial Plastic Degradation Steps: 1. Biodeterioration, 2. Biofragmentation, 3. Assimilation. [76]

**TABLE 1.10**
**Bioremediation Studies of Microplastics Using Bacteria**

| Bacterial Species | Origin of Bacteria | Type of Plastics/MPs | Reference |
|---|---|---|---|
| Bacillus sp.<br>Staphylococcus sp.<br>Pseudomonas sp. | Soil from hospital, petrol pump, and local area | 10μ and 40μ polythene | [82] |
| Ideonella saikiensis 201-F6 | Natural environment exposed to PET | Polyethylene terephthalate (PET) | [83] |
| Bacillus vallismortis<br>Psuedomonas protegens<br>Stenotrophomonas sp.<br>Paenibacillus sp. | Formulated thermophilic microbial consortia from cow dung | Low density polyethylene (LDPE) and high density polyethylene (HDPE) | [84] |
| Rhodococcus sp.<br>Shewanella sp.<br>Pseudomonas sp. | Seawater, water, and plastic and surface water sample respectively | Weathered polystyrene (PS) films | [85] |
| Arthrobacter sulfonivorans<br>Serratia plymuthica | Rhizosphere of S. viminalis growing on landfills of ash and municipal waste respectively | PET films | [86] |
| Bacillus paralicheniformis<br>B. sonorensis<br>B. licheniformis<br>B. aerius<br>B. gycinifermentans | Water sample of a plastic polluted coastal area with hot semi-arid climatic condition | Polyvinylchloride (PVC), low-density polyethylene (LDPE), and high-density polyethylene (HDPE) | [87] |
| Pseudomonas sp. ADL15<br>Rhodococcus sp. ADL36 | Fellfield soils (enriched in diesel fuel) collected from Antarctica | Polypropylene (PP) microplastics | [88] |
| Microbulbifer hydrolyticus IRE-31 | Marine pulp mill wastes rich in lignin | Low-density poly ethylene (LDPE) | [89] |

| Organism | Description | Pollutant | Ref |
|---|---|---|---|
| Bacillus cereus SEHD031MH Agromyces mediolanus PNP3 | Incubation of activated sludge at 30ºC for 60 days in growth medium supplemented with PET MPs | PET microplastics | [90] |
| Bacillus enclensis | Isolated from Biofloc samples obtained from brackish water sources | Polyethylene (PE), polypropylene (PP) and polystyrene (PS) beads | [91] |
| Vibrio sp. Cobetia sp. Pseudoalteromonas sp Ruegeria sp. Marinobacter sp. Arcobacter sp. Halodesulfovibrio sp. Pseudomonas sp. Tepidibacter sp. | Recovered from 6 samples from marine organisms (Annelida, Cnidaria, Hydrozoa, Porifera, Tunicata) from 2 marine caves | Low density polyethylene (LDPE) and bio-based PET (BPET) | [92] |

**TABLE 1.11**

**Bioremediation Studies of Microplastics Using Fungi**

| Fungal Species | Origin of Fungi | Type of Plastics/MPs | Reference |
|---|---|---|---|
| *Fusarium culmorum* | Isolated from the recycling of mixed pulper wastes in a recycled paper industry (including presence of phthalates) | Di(2-ethylhexyl) phthalate (DEHP) | [93] |
| *Zalerion maritimum* | Marine bacterium | PE microplastics (250 μm-1000 μm) | [94] |
| *Bjerkandera adusta* | Collected from mountains in South Korea | High-density polyethylene (HDPE) under lignocellulose substrate treatment | [95] |
| *Aspergillus sp.* | Isolated from marine environment | Polyethylene terephthalate (PET) | [96] |
| *Trichoderma hamatum* *Trichaptum abietinum* *Byssochlamys nivea* *B.amyloliquefaciens* | Compost sample from the municipal composting plant, black mulching film removed from agricultural soil, plastic sample removed from soil along a highway, sludge from the anaerobic digester at a wastewater treatment plant, and plastics removed from a landfill | LLDPE (70 μm) and LDPE (40 μm) films | [97] |

microorganisms mainly encompass chemical degradation, coagulation, filtration, and distillation whereas the future strategies are focused on enzymatic degradation and in-vivo degradation by microorganisms. [101]

### 1.9.4 Role of Enzymes

In biodegradation, enzymes catalyse oxidation, desulfurisation, dealkylation, deamination, dehalogenation, and other kinds of chemical reactions. Exoenzymes are secreted from microorganisms and they help to degrade polymers to monomers by binding with the substrate and digesting polymers. Fungus species secrete laccase, cutinase, and lignin peroxidase to degrade polymers whereas bacteria secrete plastic degrading enzymes like hydrolase, PETase, protease, cutinase. [102] PETase, and MHETase activity in I. sakiensis hydrolyses PET into monomers, ethylene glycol, and terephthalic acid. [83] By the genetical engineering method, secretion of thermophilic cutinase from anaerobic thermophilic bacterium *Clostridium thermocellum* was enabled and degradation of PET at up to 70°C was achieved by using the derived cutinase. [103]

By laccase and esterase activity, di(2-ethyl hexyl) phthalate (DEHP) was degraded in butanediol as an end product by fungi *Fusarium culmorum*. [93] A notable reduction (31%) in poly vinyl chloride (PVC) films was observed within four weeks of the incubation with fungal strain *Phanerocheate chrysosporium* secreting a lignin peroxidase enzyme. [104]

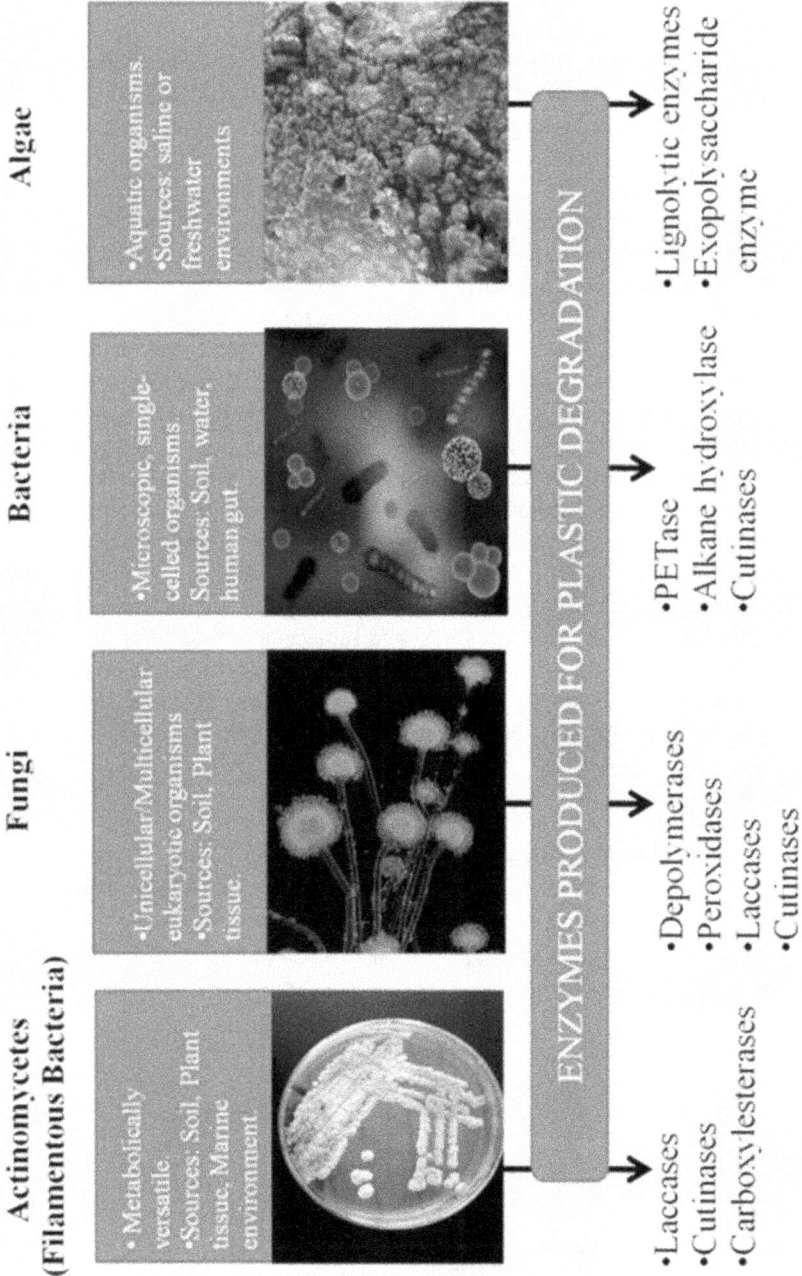

**FIGURE 1.2** Plastic degrading enzymes from microorganisms: actinomycetes, fungi, bacteria, and algae [102]

## 1.10  BIOREMEDIATION OF PHARMACEUTICAL WASTE

From the advanced medical science and pharmacology, various drugs were developed in recent years to cure rare and frequent diseases. [105] Generally, pharmaceuticals were hydrophilic in nature and were biologically active compounds which can easily absorbed by the body. So, using excessive pharmaceutical compounds can affect humans, animals, and environmental health. The derived pharmaceutical pollutants in presence of water were a long-time risk to the aquatic ecosystem even in a smaller amount. The most consumed drugs, antibiotics, anti-inflammatory drugs, anti-hypertensive drugs, beta-blockers, diuretics, etc., were detected in the influent water of wastewater treatment plants. [106] Some of the drugs were mentioned in Table 1.12.

Occurrence of pharmaceutical compounds in effluent water from wastewater treatment plants and environmental surface water bodies and metabolites of antibiotics like "oxytetracycline and clarithromycin" show the toxic effect on the water animalse like fishes, microcrustacean, green alga, cyanobacterium, [110] and rats. [111] Fungi are the microorganisms which can degrade the organic compounds that present in environment; comparatively white rot fungi are a composed and eco-friendly group that have the capability to degrading lignin, including a large number of pharmaceuticals and antibiotics that can be degraded. The steps involved for developing the device were isolation and characterisation of the novel strain that has the degradation rate and selection of fungus which can have various enzyme activates based on compatibility antagonism tests. Next, selection of a suitable matrix and establishment of the fermentation process for the production and design and construction of the container device. Finally, the evaluation test and replacement frequency and optimisation of the system. By using the fungal cocktails, as well as their immobilisation and disposal in a removable device, a possible innovation of high value for the treatment of these pollutants is represented. In conclusion, white rot fungi are used as tool for

## TABLE 1.12
### Occurrence of Pharmaceutical Compounds in Effluent Water from Wastewater Treatment Plants and Environmental Surface Water Bodies

| Drug Name | Drug Type | Site | Reference |
|---|---|---|---|
| 1. Paracetamol (acetaminophen) 2. Ibuprofen 3. Naproxen 4. Diclofenac | Anti-inflammatory analgesics | Effluent wastewater treatment plant, surface water (river), drinking water | [107] |
| 1. Metformin | Antidiabetic | Hospital effluent wastewater treatment plant | [108] |
| 1. Atenolol 2. Fenofibrate | Anti-hypertensives | Effluent wastewater treatment plant | [107] |
| 1. Clarithromycin 2. Erythromycin 3. Ofloxacin 4. Levofloxacin 5. Ciprofloxacin | Antibiotics | Surface water (river), effluent wastewater treatment plant | [109, 107] |

the bioremediation of the contaminants of pharmaceutical origin; they are excellent biological agents to include in wastewater treatment process. Here the removable device used is more efficient bioremediation of emerging pollutants and a more efficient globe water treatment. [112]

## 1.11   ADVANTAGES AND LIMITATIONS

Bioremediation being an economic and sustainable approach has several advantages. It can be performed on site and degrades complex harmful substances into its less toxic forms. Yet bioremediation is a newly evolved technology and hence has a few shortcomings which constrict its application on a global scale. It is only applicable for biodegradable substances. Though it demands less manpower, it requires more skill to build an efficient system. Moreover, most of the existing studies conclude the efficiency of this type of remediation while testing in a synthetic or lab-made setup. This is the reason why real-time data is required to identify the practical problems and to explore feasible solutions for it. Advanced techniques such as microbial fuel cells should be encouraged, however, it is an expensive method. Hence, more cost-effective ways must be identified.

## 1.12   CONCLUSION

Bioremediation is not only an economic approach but also a sustainable method of contaminant removal from the environment. Despite several limitations and challenges, it is indeed one of the best and least explored areas which if implemented can bring a global revolution. Novel tools of biotechnology and genetic engineering empower such a process. Thus, bioremediation can be developed effectively on an industrial level to combat the removal of various persistent, toxic chemicals in our environment. This approach shows the way for future pollution control methods.

## ACKNOWLEDGMENTS

VJ and SS would like to acknowledge DST (SEED div), Government of India for the financial support received from DST SEED Div (SYST) project (SP/YO/2019/1283) during writing of this chapter.

## REFERENCES

[1]   Yagub, M. T., Sen, T. K., Afroze, S. and Ang, H., 2014. Dye and its removal from aqueous solution by adsorption: A review. *Adv. Colloid Interface Sci.*, *209*, pp. 172–184.
[2]   Yogalakshmi, K. N., Das, A., Rani, G., Jaswal, V. and Randhawa, J. S., 2020. Nanobioremediation: A new age technology for the treatment of dyes in textile effluents. In *Bioremediation of Industrial Waste for Environmental Safety*. Springer, Singapore, pp. 313–347.
[3]   Rauf, M. A. and Salman Ashraf, S., 2012. Survey of recent trends in biochemically assisted degradation of dyes. *Chem. Eng. J.*, *209*, pp. 520–530.
[4]   Forgacs, E., Cserháti, T. and Oros, G. 2004. Removal of synthetic dyes from wastewaters: A review. *Environ. Int.*, *30*(7), pp. 953–971.
[5]   Azubuike, C. C., Chikere, C. B. and Okpokwasili, G. C., 2016. Bioremediation techniques–classification based on site of application: Principles, advantages, limitations and prospects. *World J. Microbiol. Biotechnol.*, *32*(11), pp. 1–18.

[6]   Philp, J. C. and Atlas, R. M., 2005. Bioremediation of contaminated soils and aqui-
      fers. In *Bioremediation: Applied Microbial Solutions for Real-World Environmental
      Cleanup*, ASM Press, Washington, DC, pp. 139–236.

[7]   Whelan, M. J., Coulon, F., Hince, G., Rayner, J., McWatters, R., Spedding, T. and
      Snape, I., 2015. Fate and transport of petroleum hydrocarbons in engineered biopiles in
      polar regions. *Chemosphere, 131*, pp. 232–240.

[8]   Sanscartier, D., Zeeb, B., Koch, I. and Reimer, K., 2009. Bioremediation of diesel-
      contaminated soil by heated and humidified biopile system in cold climates. *Cold Reg.
      Sci. Technol, 55*(1), pp. 167–173.

[9]   Hobson, A. M., Frederickson, J. and Dise, N. B., 2005. CH4 and N2O from mechanically
      turned windrow and vermicomposting systems following in-vessel pre-treatment. *Waste
      Manage, 25*(4), pp. 345–352.

[10]  Affat, S. S., 2021. Classifications, advantages, disadvantages, toxicity effects of natural
      and synthetic dyes: A review. *University of Thi-Qar J. Sci., 8*(1), pp. 130–135.

[11]  Kadolph, S., 2008. Natural dyes: A traditional craft experiencing new attention. *Delta
      Kappa Gamma Bulletin, 75*(1), p. 14.

[12]  Choudhury, A. K. R., 2018. Eco-friendly dyes and dyeing. *Adv. Mat. Tech. Env., 2*,
      pp. 145–176.

[13]  Mathur, N., Bhatnagar, P. and Bakre, P., 2006. Assessing mutagenicity of textile dyes
      from Pali(Rajasthan) using Ames bioassay. *Appl. Ecol. Env. Res., 4*(1), pp. 111–118.

[14]  Giovanella, P., Vieira, G. A., Ramos Otero, I. V., Pais Pellizzer, E., de Jesus Fontes, B.
      and Sette, L. D., 2020. Metal and organic pollutants bioremediation by extremophile
      microorganisms. *J. Hazard. Mater., 382*, p. 121024.

[15]  Ajaz, M., Shakeel, S. and Rehman, A., 2019. Microbial use for azo dye degradation—a
      strategy for dye bioremediation. *Int. Microbiol., 23*(2), pp. 149–159.

[16]  Paz, A., Carballo, J., Pérez, M. J. and Domínguez, J. M., 2017. Biological treatment of
      model dyes and textile wastewaters. *Chemosphere, 181*, pp. 168–177.

[17]  Patel, D. K., Tipre, D. R. and Dave, S. R., 2017. Enzyme mediated bacterial biotrans-
      formation and reduction in toxicity of 1:2 chromium complex AB193 and AB194 dyes.
      *J. Taiwan Inst. Chem. Eng., 77*, pp. 1–9.

[18]  Sofu, A., 2018. Investigation of dye removal with isolated biomasses from whey waste-
      water. *Int. J. Environ. Sci. Technol., 16*(1), pp. 71–78.

[19]  Neifar, M., Chouchane, H., Mahjoubi, M., Jaouani, A. and Cherif, A., 2016. Pseu-
      domonas extremorientalis BU118: A new salt-tolerant laccase-secreting bacterium
      with biotechnological potential in textile azo dye decolourization. *3 Biotech, 6*(1),
      pp. 1–9.

[20]  Blánquez, A., Rodríguez, J., Brissos, V., Mendes, S., Martins, L. O., Ball, A. S., Arias,
      M. E. and Hernández, M., 2019. Decolorization and detoxification of textile dyes using
      a versatile Streptomyces laccase-natural mediator system. *Saudi J. Biol. Sci., 26*(5),
      pp. 913–920.

[21]  Karatay, S. E., Kılıç, N. K. and Dönmez, G., 2015. Removal of Remazol Blue by azo-
      reductase from newly isolated bacteria. *Ecol. Eng., 84*, pp. 301–304.

[22]  El Bouraie, M. and El Din, W. S., 2016. Biodegradation of Reactive Black 5 by Aeromo-
      nas hydrophila strain isolated from dye-contaminated textile wastewater. *Sustainable
      Environ. Res., 26*(5), pp. 209–216.

[23]  Guadie, A., Gessesse, A. and Xia, S., 2018. Halomonas sp. strain A55, a novel dye
      decolorizing bacterium from dye-uncontaminated Rift Valley Soda lake. *Chemosphere,
      206*, pp. 59–69.

[24]  Xie, X. H., Zheng, X. L., Yu, C. Z., Zhang, Q. Y., Wang, Y. Q., Cong, J. H., Liu, N., He,
      Z. J., Yang, B. and Liu, J. S., 2019. High-efficient biodegradation of refractory dye by
      a new bacterial flora DDMY1 under different conditions. *Int. J. Environ. Sci. Technol.,
      17*(3), pp. 1491–150.

[25] Ihsanullah, I., Jamal, A., Ilyas, M., Zubair, M., Khan, G. and Atieh, M. A., 2020. Bioremediation of dyes: Current status and prospects. *J. Water Process. Eng.*, *38*, p. 101680.

[26] Neoh, C. H., Lam, C. Y., Lim, C. K., Yahya, A., Bay, H. H., Ibrahim, Z. and Noor, Z. Z., 2015. Biodecolorization of recalcitrant dye as the sole sourceof nutrition using Curvularia clavata NZ2 and decolorization ability of its crude enzymes. *Environ. Sci. Pollut. Res.*, *22*(15), pp. 11669–11678.

[27] Pandi, A., Marichetti Kuppuswami, G., Numbi Ramudu, K. and Palanivel, S., 2019. A sustainable approach for degradation of leather dyes by a new fungal laccase. *J. Cleaner Prod.*, *211*, pp. 590–597.

[28] Mahmoud, M. S., Mostafa, M. K., Mohamed, S. A., Sobhy, N. A. and Nasr, M., 2017. Bioremediation of red azo dye from aqueous solutions by a spergillus niger strain isolated from textile wastewater. *J. Environ. Chem. Eng.*, *5*(1), pp. 547–554.

[29] Arunprasath, T., Sudalai, S., Meenatchi, R., Jeyavishnu, K. and Arumugam, A., 2019. Biodegradation of triphenylmethane dye malachite green by a newly isolated fungus strain. *Biocatal. Agric. Biotechnol.*, *17*, pp. 672–679.

[30] Li, S., Huang, J., Mao, J., Zhang, L., He, C., Chen, G., Parkin, I. P. and Lai, Y., 2019. *In vivo* and *in vitro* efficient textile wastewater remediation by *Aspergillus niger* biosorbent. *Nanoscale Adv.*, *1*(1), pp. 168–176.

[31] Kulkarni, A. N., Watharkar, A. D., Rane, N. R., Jeon, B. H. and Govindwar, S. P., 2018. Decolorization and detoxification of dye mixture and textile effluent by lichen Dermatocarpon vellereceum in fixed bed upflow bioreactor with subsequent oxidative stress study. *Ecotoxicol. Environ. Saf.*, *148*, pp. 17–25.

[32] Afshariani, F. and Roosta, A., 2019. Experimental study and mathematical modeling of biosorption of methylene blue from aqueous solution in a packed bed of microalgae Scenedesmus. *J. Cleaner Prod.*, *225*, pp. 133–142.

[33] Tan, L., He, M., Song, L., Fu, X. and Shi, S., 2016. Aerobic decolorization, degradation and detoxification of azo dyes by a newly isolated salt-tolerant yeast Scheffersomyces spartinae TLHS-SF1. *Bioresour. Technol.*, *203*, pp. 287–294.

[34] Martorell, M. M., Pajot, H. F. and Figueroa, L. I. D., 2017. Biological degradation of Reactive Black 5 dye by yeast Trichosporon akiyoshidainum. *J. Environ. Chem. Eng.*, *5*(6), pp. 5987–5993.

[35] Nguyen, T. A., Fu, C. C. and Juang, R. S., 2016. Effective removal of sulfur dyes from water by biosorption and subsequent immobilized laccase degradation on crosslinked chitosan beads. *Chem. Eng.*, *304*, pp. 313–324.

[36] Teerapatsakul, C., Parra, R., Keshavarz, T. and Chitradon, L., 2017. Repeated batch for dye degradation in an airlift bioreactor by laccase entrapped in copper alginate. *Int. Biodeterior. Biodegrad.*, *120*, pp. 52–57.

[37] Buscio, V., García-Jiménez, M., Vilaseca, M., López-Grimau, V., Crespi, M. and Gutiérrez-Bouzán, C., 2016. Reuse of textile dyeing effluents treated with coupled nanofiltration and electrochemical processes. *Materials*, *9*(6), p. 490.

[38] Bilal, M., Iqbal, H. M., Hussain Shah, S. Z., Hu, H., Wang, W. and Zhang, X., 2016. Horseradish peroxidase-assisted approach to decolorize and detoxify dye pollutants in a packed bed bioreactor. *J. Environ. Manage.*, *183*, pp. 836–842.

[39] Bento, R. M., Almeida, M. R., Bharmoria, P., Freire, M. G. and Tavares, A. P., 2020. Improvements in the enzymatic degradation of textile dyes using ionic-liquid-based surfactants. *Sep. Purif. Technol.*, *235*, p. 116191

[40] Yang, X., Zheng, J., Lu, Y. and Jia, R., 2016. Degradation and detoxification of the triphenylmethane dye malachite green catalyzed by crude manganese peroxidase from Irpex lacteus F17. *Environ. Sci. Pollut. Res.*, *23*(10), pp. 9585–9597.

[41] Nouren, S., Bhatti, H. N., Iqbal, M., Bibi, I., Kamal, S., Sadaf, S., Sultan, M., Kausar, A. and Safa, Y., 2017. By-product identification and phytotoxicity of biodegraded Direct Yellow 4 dye. *Chemosphere*, *169*, pp. 474–484.

[42]   Yadav, K. K., Gupta, N., Kumar, V. and Singh, J. K., 2017. Bioremediation of heavy metals from contaminated sites using potential species: A review. *Indian J. Environ. Prot*, *37*(1), p. 65.

[43]   Mythili, K. and Karthikeyan, B., 2011. Bioremediation of chromium [Cr (VI)] in tannery effluent using Bacillus spp. and Staphylococcus spp. *Int J Pharm Biol Arch*, *2*(5), pp. 1460–1463.

[44]   Manjengwa, F., Nhiwatiwa, T., Nyakudya, E. and Banda, P., 2019. Fish from a polluted lake (Lake Chivero, Zimbabwe): A food safety issue of concern. *Food Qual. Saf.*, *3*(3), pp. 157–167.

[45]   Rajendran, P., Muthukrishnan, J. and Gunasekaran, P., 2003. Microbes in heavy metal remediation. *Indian J. Exp. Biol.*, *41*(9), pp. 935–944.

[46]   Ahemad, M., 2012. Implications of bacterial resistance against heavy metals in bioremediation: A review. *J IIOAB*, *3*(3).

[47]   Dey, U., Chatterjee, S. and Mondal, N. K., 2016. Isolation and characterization of arsenic-resistant bacteria and possible application in bioremediation. *Biotechnol. Rep.*, *10*, pp. 1–7.

[48]   He, Y., Gong, Y., Su, Y., Zhang, Y. and Zhou, X., 2019. Bioremediation of Cr (VI) contaminated groundwater by Geobacter sulfurreducens: Environmental factors and electron transfer flow studies. *Chemosphere*, *221*, pp. 793–801.

[49]   Chellaiah, E. R., 2018. Cadmium (heavy metals) bioremediation by Pseudomonas aeruginosa: A minireview. *Appl. Water Sci.*, *8*(6), pp. 1–10.

[50]   Tariq, A., Ullah, U., Asif, M. and Sadiq, I., 2018. Biosorption of arsenic through bacteria isolated from Pakistan. *Int. Microbiol.*, *22*(1), pp. 59–68.

[51]   Gao, R., Wang, Y., Zhang, Y., Tong, J. and Dai, W., 2017. Cobalt(II) bioaccumulation and distribution in *Rhodopseudomonas palustris*. *Biotechnol. Biotechnol. Equip.*, *31*(3), pp. 527–534.

[52]   Purvis, O. W. and Halls, C., 1996. A review of lichens in metal-enriched environments. *The Lichenologist*, *28*(6), pp. 571–601.

[53]   Cárdenas-González, J. F., Acosta-Rodriguez, I., Téran-Figueroa, Y. and Rodriguez-Perez, A. S., 2017. Bioremoval of arsenic (V) from aqueous solutions by chemically modified fungal biomass. *3 Biotech*, *7*(3), pp. 1–6.

[54]   Al-Hares, H. S., 2017. Single and binary biosorption isotherms of different heavy metal ions using fungal waste biomass. *Al-Nahrain J. Eng. Sci.*, *20*(3), pp. 673–684.

[55]   Bahobil, A., Bayoumi, R., Atta, H., M. and Sehrawey, E., 2017. Fungal biosorption for cadmium and mercury heavy metal ions isolated from some polluted localities in KSA. *Int. J. Curr. Microbiol. Appl. Sci.*, *6*(6), pp. 2138–2154.

[56]   Bano, A., Hussain, J., Akbar, A., Mehmood, K., Anwar, M., Hasni, M. S., Ullah, S., Sajid, S. and Ali, I., 2018. Biosorption of heavy metals by obligate halophilic fungi. *Chemosphere*, *199*, pp. 218–222.

[57]   Manguilimotan, L. C. and Bitacura, J. G., 2018. Biosorption of cadmium by filamentous fungi isolated from coastal water and sediments. *J. Toxicol.*, *2018*, 1–6.

[58]   Frutos, I., García-Delgado, C., Gárate, A. and Eymar, E., 2016. Biosorption of heavy metals by organic carbon from spent mushroom substrates and their raw materials. *Int. J. Environ. Sci. Technol.*, *13*(11), pp. 2713–2720.

[59]   Dwivedi, S., 2012. Bioremediation of heavy metal by algae: Current and future perspective. *J. Adv. Lab. Res. Biol.*, *3*(3), pp. 195–199.

[60]   Goher, M. E., AM, A. E. M., Abdel-Satar, A. M., Ali, M. H., Hussian, A. E. and Napiórkowska-Krzebietke, A. 2016. Biosorption of some toxic metals from aqueous solution using non-living algal cells of Chlorella vulgaris. *J. Elementology*, *21*(3), pp. 703–714.

[61]   Megharaj, M., Ramakrishnan, B., Venkateswarlu, K., Sethunathan, N. and Naidu, R., 2011. Bioremediation approaches for organic pollutants: a critical perspective. *Environ Int.*, *37*(8), pp. 1362–1375.

[62] Uthayakumar, H., Radhakrishnan, P., Shanmugam, K. and Kushwaha, O. S., 2022. Growth of MWCNTs from Azadirachta indica oil for optimization of chromium (VI) removal efficiency using machine learning approach. *Environ Sci Pollut. Res.*, *29*(23), pp. 34841–34860.

[63] Watanabe, K., 2001. Microorganisms relevant to bioremediation. *Curr. Opin. Biotechnol.*, *12*(3), pp. 237–241.

[64] Karigar, C. S. and Rao, S. S., 2011. Role of microbial enzymes in the bioremediation of pollutants: A review. *Enzyme Res.*, *2011*.

[65] Haripriyan, U., Gopinath, K. P., Arun, J. and Govarthanan, M., 2022. Bioremediation of organic pollutants: a mini review on current and critical strategies for wastewater treatment. *Arch. Microbiol.*, *204*(5), pp. 1–9.

[66] Rempel, A., Gutkoski, J. P., Nazari, M. T., Biolchi, G. N., Cavanhi, V. A. F., Treichel, H. and Colla, L. M., 2021. Current advances in microalgae-based bioremediation and other technologies for emerging contaminants treatment. *Sci. Total Environ.*, *772*, p. 144918.

[67] Gattullo, C. E., Bährs, H., Steinberg, C. E. and Loffredo, E., 2012. Removal of bisphenol A by the freshwater green alga Monoraphidium braunii and the role of natural organic matter. *Sci. Total Environ.*, *416*, pp. 501–506.

[68] Munoz, R., Guieysse, B. and Mattiasson, B., 2003. Phenanthrene biodegradation by an algal-bacterial consortium in two-phase partitioning bioreactors. *Appl. Microbiol. Biotechnol.*, *61*(3), pp. 261–267.

[69] Karaismailoglu, M. C., 2015. Investigation of the potential toxic effects of prometryne herbicide on Allium cepa root tip cells with mitotic activity, chromosome aberration, micronucleus frequency, nuclear DNA amount and comet assay. *Caryologia*, *68*(4), pp. 323–329.

[70] Dosnon-Olette, R., Trotel-Aziz, P., Couderchet, M. and Eullaffroy, P., 2010. Fungicides and herbicide removal in Scenedesmus cell suspensions. *Chemosphere*, *79*(2), pp. 117–123.

[71] Semple, K. T., Cain, R. B. and Schmidt, S., 1999. Biodegradation of aromatic compounds by microalgae. *FEMS Microbiol. Lett.*, *170*(2), pp. 291–300.

[72] GESAMP Joint Group of Experts on the Scientific Aspects of Marine Environmental Protection., 2016. Sources, fate and effects of microplastics in the marine environment: part 2 of a global assessment.(IMO, FAO/UNESCO-IOC/UNIDO/WMO/IAEA/UN/UNEP/UNDP). In: Kershaw, P. J. (Ed.), Rep. Stud. GESAMP No. 90 (96 pp). *Reports and Studies GESAMP, No. 93, 96 P.*, *93*.

[73] MFRC-GMIT., 2021. *Microplastics in the Marine Environment Report_Final_15092021*. Norwegian Institute for Water Research, Oslo.

[74] UNEP., 2016. *Recommended citation: Acknowledgements: Marine Plastic Debris and Microplastics—Global Lessons and Research to Inspire Action and Guide Policy Change*. United Nations Environment Programme, Nairobi.

[75] Avio, C. G., Gorbi, S. and Regoli, F., 2017. Plastics and microplastics in the oceans: From emerging pollutants to emerged threat. *Mar. Environ. Res.*, *128*, pp. 2–11.

[76] Cunningham, E. M., Mundye, A., Kregting, L., Dick, J. T., Crump, A., Riddell, G. and Arnott, G., 2021. Animal contests and microplastics: evidence of disrupted behaviour in hermit crabs Pagurus bernhardus. *R. Soc. Open Sci.*, *8*(10), p. 211089.

[77] Revell, L. E., Kuma, P., Le Ru, E. C., Somerville, W. R. and Gaw, S., 2021. Direct radiative effects of airborne microplastics. *Nature*, *598*(7881), pp. 462–467.

[78] Zeenat, Elahi, A., Bukhari, D. A., Shamim, S. and Rehman, A., 2021. Plastics degradation by microbes: A sustainable approach. *Journal of King Saud University—Science*, *33*(6), p. 101538.

[79] Shahnawaz, M., Sangale, M. K. and Ade, A. B., 2019. Bioremediation Technology for Plastic Waste. In *Bioremediation Technology for Plastic Waste*. Springer Singapore, Singapore, pp. 978–981.

[80] Sheth, M. U., Kwartler, S. K., Schmaltz, E. R., Hoskinson, S. M., Martz, E. J., Dunphy-Daly, M. M., Schultz, T. F., Read, A. J., Eward, W. C. and Somarelli, J. A., 2019.

Bioengineering a future free of marine plastic waste. *Front. Mar. Sci.*, *6*(October), pp. 1–10.

[81] Hadian-Ghazvini, S., Hooriabad Saboor, F. and Safaee Ardekani, L., 2022. Bioremediation techniques for microplastics removal. *Microplastics Pollution in Aquatic Media* (Springer), pp. 327–377.

[82] Singh, G., Singh, A. K. and Bhatt, K., 2016. Biodegradation of polyethylene by bacteria isolated from soil. *Int. J. Res. Dev. Pharm. L. Sci*, *5*(2), pp. 2056–2062.

[83] Yoshida, S., Hiraga, K., Takehana, T., Taniguchi, I., Yamaji, H., Maeda, Y., Toyohara, K., Miyamoto, K., Kimura, Y. and Oda, K., 2016. Response to comment on "A bacterium that degrades and assimilates poly (ethylene terephthalate)". *Science*, *353*(6301), 759–759.

[84] Skariyachan, S., Setlur, A. S., Naik, S. Y., Naik, A. A., Usharani, M. and Vasist, K. S., 2017. Enhanced biodegradation of low and high-density polyethylene by novel bacterial consortia formulated from plastic-contaminated cow dung under thermophilic conditions. *Environ. Sci. Pollut. Res.*, *24*(9), pp. 8443–8457.

[85] Syranidou, E., Karkanorachaki, K., Amorotti, F., Franchini, M., Repouskou, E., Kaliva, M., Vamvakaki, M., Kolvenbach, B., Fava, F., Corvini, P. F. X. and Kalogerakis, N., 2017. Biodegradation of weathered polystyrene films in seawater microcosms. *Sci. Rep.*, *7*(1), pp. 1–12.

[86] Janczak, K., Hrynkiewicz, K., Znajewska, Z. and Dąbrowska, G., 2018. Use of rhizosphere microorganisms in the biodegradation of PLA and PET polymers in compost soil. *Int. Biodeterior. Biodegrad.*, *130*(November 2017), pp. 65–75.

[87] Kumari, A., Chaudhary, D. R. and Jha, B., 2019. Destabilization of polyethylene and polyvinylchloride structure by marine bacterial strain. *Environ. Sci. Pollut. Res.*, *26*(2), pp. 1507–1516.

[88] Habib, S., Iruthayam, A., Shukor, M. Y. A., Alias, S. A., Smykla, J. and Yasid, N. A., 2020. Biodeterioration of untreated polypropylene microplastic particles by antarctic bacteria. *Polymers*, *12*(11), pp. 1–12.

[89] Li, Z., Wei, R., Gao, M., Ren, Y., Yu, B., Nie, K., Xu, H. and Liu, L., 2020. Biodegradation of low-density polyethylene by Microbulbifer hydrolyticus IRE-31. *J. Environ. Manage.*, *263*, p. 110402.

[90] Torena, P., Alvarez-Cuenca, M. and Reza, M., 2021. Biodegradation of polyethylene terephthalate microplastics by bacterial communities from activated sludge. *Can. J. Chem. Eng.*, *99*, pp. S69–S82.

[91] Ahmad Shukri, Z. N., Che Engku Chik, C. E. N., Hossain, S., Othman, R., Endut, A., Lananan, F., Terkula, I. B., Kamaruzzan, A. S., Abdul Rahim, A. I., Draman, A. S. and Kasan, N. A. 2022. A novel study on the effectiveness of bioflocculant-producing bacteria Bacillus enclensis, isolated from biofloc-based system as a biodegrader in microplastic pollution. *Chemosphere*, *308*(Pt2), pp. 136410.

[92] de Villalobos, N. F., Costa, M. C. and Marín-Beltrán, I., 2022. A community of marine bacteria with potential to biodegrade petroleum-based and biobased microplastics. *Mar. Pollut. Bull.*, *185*, p. 114251.

[93] Ahuactzin-Pérez, M., Tlecuitl-Beristain, S., García-Dávila, J., González-Pérez, M., Gutiérrez-Ruíz, M. C. and Sánchez, C., 2016. Degradation of di(2-ethyl hexyl) phthalate by Fusarium culmorum: Kinetics, enzymatic activities and biodegradation pathway based on quantum chemical modelingpathway based on quantum chemical modeling. *Sci. Total Environ.*, *566*, pp. 1186–1193.

[94] Paço, A., Duarte, K., da Costa, J. P., Santos, P. S. M., Pereira, R., Pereira, M. E., Freitas, A. C., Duarte, A. C. and Rocha-Santos, T. A. P., 2017. Biodegradation of polyethylene microplastics by the marine fungus Zalerion maritimum. *Sci. Total Environ.*, *586*, pp. 10–15.

[95] Kang, B. R., Kim, S. Bin, Song, H. A. and Lee, T. K., 2019. Accelerating the biodegradation of high-density polyethylene (HDPE) using Bjerkandera adusta TBB-03 and lignocellulose substrates. *Microorganisms*, *7*(9), p. 304.

[96]    Sarkhel, R., Sengupta, S., Das, P. and Bhowal, A., 2020. Comparative biodegradation study of polymer from plastic bottle waste using novel isolated bacteria and fungi from marine source. *J. Polym. Res.*, *27*(1), pp. 1–8.

[97]    Malachová, K., Novotný, Č., Adamus, G., Lotti, N., Rybková, Z., Soccio, M., Šlosarčíková, P., Verney, V. and Fava, F., 2020. Ability of Trichoderma hamatum isolated from plastics-polluted environments to attack petroleum-based, synthetic polymer films. *Processes*, *8*(4), p. 467.

[98]    Vimal Kumar, R., Kanna, G. R. and Elumalai, S., 2017. Biodegradation of Polyethylene by Green Photosynthetic Microalgae. *J. Biorem. Biodegrad.*, *08*(01), pp. 1–8.

[99]    Moog, D., Schmitt, J., Senger, J., Zarzycki, J., Rexer, K. H., Linne, U., Erb, T. and Maier, U. G., 2019. Using a marine microalga as a chassis for polyethylene terephthalate (PET) degradation. *Microb. Cell Fact.*, *18*(1), pp. 1–15.

[100]   Kim, J. W., Park, S. Bin, Tran, Q. G., Cho, D. H., Choi, D. Y., Lee, Y. J. and Kim, H. S., 2020. Functional expression of polyethylene terephthalate-degrading enzyme (PETase) in green microalgae. *Microb. Cell Fact.*, *19*(1), pp. 1–9.

[101]   Barone, G. D., Ferizović, D., Biundo, A. and Lindblad, P. (2020). Hints at the applicability of microalgae and cyanobacteria for the biodegradation of plastics. *Sustainability (Switzerland)*, *12*(24), pp. 1–15.

[102]   Kaushal, J., Khatri, M. and Arya, S. K., 2021. Recent insight into enzymatic degradation of plastics prevalent in the environment: A mini—review. *J. Cleaner Prod.*, *2*, p. 100083.

[103]   Yan, F., Wei, R., Cui, Q., Bornscheuer, U. T. and Liu, Y. J., 2021. Thermophilic whole-cell degradation of polyethylene terephthalate using engineered Clostridium thermocellum. *Microb. Biotechnol.*, *14*(2), pp. 374–385.

[104]   Khatoon, N., Jamal, A. and Ali, M. I., 2019. Lignin peroxidase isoenzyme: a novel approach to biodegrade the toxic synthetic polymer waste. *Environ. Technol.*, *40*(11), pp. 1366–1375.

[105]   Podolsky, S. H., 2018. The evolving response to antibiotic resistance (1945–2018). *Palgrave Communications*, *4*(1), pp. 1–8.

[106]   Verlicchi, P., Al Aukidy, M. and Zambello, E., 2012. Occurrence of pharmaceutical compounds in urban wastewater: Removal, mass load and environmental risk after asecondary treatment-a review. *Sci. Total Environ.*, *429*, pp. 123–155.

[107]   Biel-Maeso, M., Corada-Fernández, C. and Lara-Martín, P. A., 2018. Monitoring the occurrence of pharmaceuticals in soils irrigated with reclaimed wastewater. *Environ. Pollut.*, *235*, pp. 312–321.

[108]   Papageorgiou, M., Zioris, I., Danis, T., Bikiaris, D. and Lambropoulou, D., 2019. Comprehensive investigation of a wide range of pharmaceuticals and personal care products in urban and hospital wastewaters in Greece. *Sci. Total Environ.*, *694*, p. 133565.

[109]   Mahmood, A. R., Al-Haideri, H. H. and Hassan, F. M., 2019. Detection of antibiotics in drinking water treatment plants in Baghdad City, Iraq. *Adv. Public Health*, *2019*, pp. 1–10.

[110]   Baumann, M., Weiss, K., Maletzki, D., Schüssler, W., Schudoma, D., Kopf, W. and Kühnen, U., 2015. Aquatic toxicity of the macrolide antibiotic clarithromycin and its metabolites. *Chemosphere*, *120*, pp. 192–198.

[111]   Tan, L., He, M., Song, L., Fu, X. and Shi, S., 2016. Aerobic decolorization, degradation and detoxification of azo dyes by a newly isolated salt-tolerant yeast Scheffersomyces spartinae TLHS-SF1. *Bioresour. Technol.*, *203*, pp. 287–294.

[112]   Akerman-Sanchez, G. and Rojas-Jimenez, K., 2021. Fungi for the bioremediation of pharmaceutical-derived pollutants: A bioengineering approach to water treatment. *Environ. Adv.*, *4*, p. 100071.

# 2 Green Synthesis of Iron Oxide Nanoparticles and Its Application in Water Treatment

*Alli Malar Harikrishnan, Zaira Zaman Chowdhury, Masud Rana, Ahmed Elsayid Ali, Ajita Mitra, Rahman Faizur Rafique and Rafie Bin Johan*

## 2.1 INTRODUCTION

The field of nanotechnology is one of the most productive and rapidly expanding subfields of modern research, which has enjoyed a great deal of achievement as a result of advancements made in contemporary technology. Nanoparticles are incredibly fascinating because they exist in a transitional state between the atomic or molecular structures of bulk matter and nanoparticles themselves. Nanoparticles are a specific kind of material that can be distinguished from other particles by their sizes (which typically fall within the range of one to one hundred nm), framework, physico-chemical properties including magnetic behaviour, electrical, mechanical, thermal, catalytic and light-scattering capabilities [1, 2]. Nanoparticles are so small and have such a large surface active area for their size that their attributes and activity are almost entirely determined by their sizes [3]. At the nano-level, this situation frequently takes on a different form than it does for bulk materials, which have consistent chemical and physical properties irrespective of their size. When examined on a nanoscale, a number of bulk materials have shown remarkable properties [4]. In terms of the nanoparticles' structures, a high degree of intricacy may be seen. They are made up of three layers, which are the surface exterior layer, the shell, and the core respectively. The functional groups, such as metallic ions, surfactant, molecules, and polymers, which make up each layer are distinct from one another. In most cases, the nanoparticles are used to represent the core [5]. Nanoparticles have their own unique characteristics, which must be tuned during the synthesis pathway. These characteristics include dimension, shape, content, and the hierarchical framework [6].

Many different metal oxides (MO), such as copper oxide, tungsten (divalent, trivalent) oxide, nickel oxide, zinc oxide, titanium dioxide, silver oxide, tin oxide, iron oxide, silicon oxide, and gold oxide have a comprehensive range of usages in a variety of fields, including the environment (identification of toxins and contaminants, environmental cleanup, water treatment, photo degradation), catalysis, paint

DOI: 10.1201/9781003342830-2

and textile industries, electronic industries (optical limiting devices, batteries, gas sensors), and mechanical industries (nano-generators) [7].

In the process of manufacturing IONPs with desirable properties based on their surface functional groups, a variety of procedures, including chemical, physical, and biological processes, have been utilised. These procedures are depicted in Figure 2.1. The most recent bio-synthesis procedures, the characterisation of the nanoparticles, and their applications are discussed in depth in [8]. Approaches from the top down as well as the bottom up might be utilised in the production of nanoparticles. The top-down methodology takes into account only the physically occurring processes, whereas the bottom-up methodology takes into account the biological and chemical processes [8].

One of the most biocompatible NPs out of these MO is iron oxide (IONPs) because it possesses splendid vanishingly small physical features such as super para magnetism, stiffness in fluid medium, low susceptibility to oxidation, long plasma half-lives, and adaptable surface composition. These IONPs have a diverse range of applications in environmental legislation such as degradation of organic compounds, adsorption of heavy metals and dyes, nutrition associated operations, biomedicine (drug delivery, magnetic particle imaging (MPI), magnetic resonance imaging (MRI), cancer treatment, etc.) [9, 10] based on their forms as exhibited by Figure 2.2. It has antibacterial activity against pathogens.

In addition, the use of chemical-based methods necessitates the employment of solvents such as hydrazine, sodium borohydride, and sodium dodecyl sulphate and potassium tartrate. Each of these substances contributes to the production of toxic waste flows, which makes them detrimental to the natural environment. In addition, the physical generation process involves grinding, milling, and electroporation. All of these processes are expensive per hour because of the extensive amount of energy consumption. The extremely low output yields of this method deliver another significant drawback of using it [11].

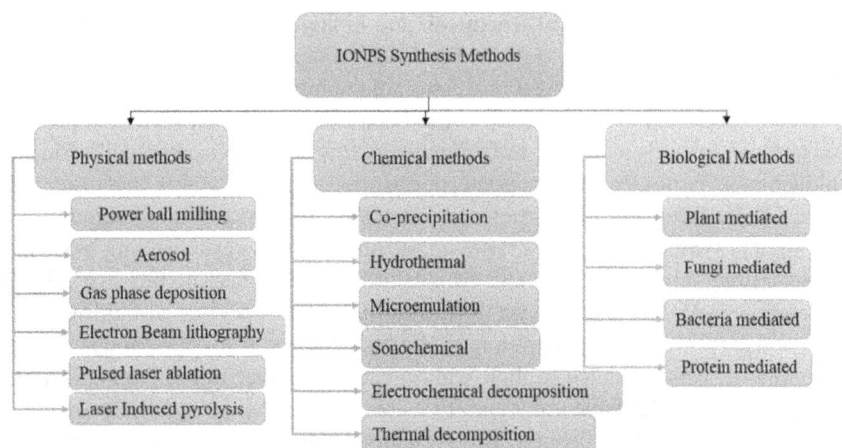

FIGURE 2.1  Different Methods for Synthesis of IONPs Nanoparticles.

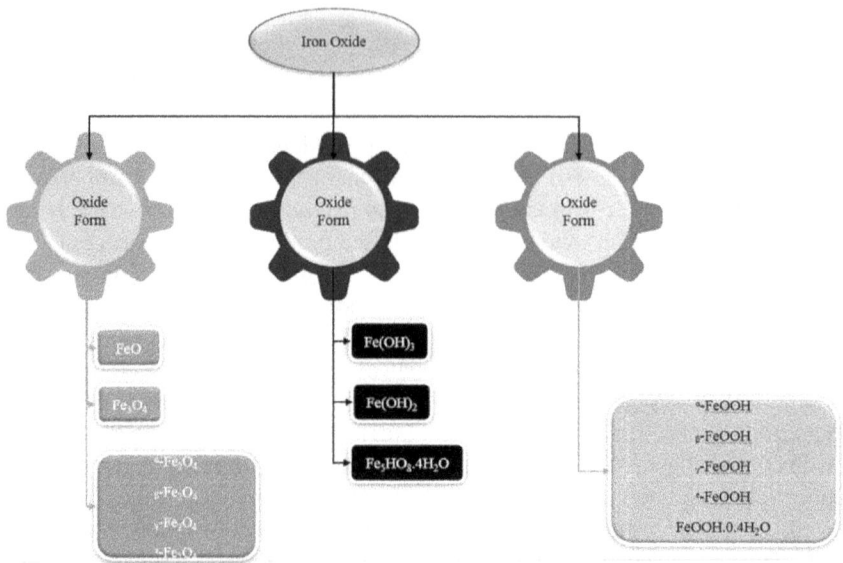

**FIGURE 2.2** Different Forms of IONPs Nanoparticles.

In more recent times, IONPs have also been described as "Nanozymes" due to the inherent enzyme-like functions they exhibit. They have catalytic properties that are comparable to those of a number of other oxidases, such as catalase, peroxidases, superoxide dismutase, and sulfite oxidase [12]. These particles have the remarkable ability to form bonds with a wide variety of biological components, including enzymes, peptides, nucleic acids, triglycerides, fatty acids, and various metabolites. When compared to their un-immobilised equivalents, molecules that have been immobilised on nanoparticles exhibit excellent properties and a top level of reusability [13]. The production of IONPs has been accomplished by a variety of approaches, including physical, chemical, and biological processes. The processes of flow-injection, co-precipitation, micro-emulsion methods, reversed micelles, sol-gel production, and hydrothermal processes are the most common types of reactions involved in chemical synthesis [14]. It has been established that there is a pressing need to move towards a process of fabrication that is more environmentally friendly, healthier, more biologically appropriate, consistent, financially viable, and faster [15]. Figure 2.3 illustrates the advantages of using IONPs for versatile application.

Biological synthesis of IONPs relies on the utilisation of various living biota such as plants, fungi, algae, bacteria, and virus in order to complete the process [16, 17]. This organic synthesis is reliant on the utilisation of the supercritical fluid, which is water, and it results in the generation of NPs that are free from poisonous chemical contaminants, which has led to their widespread acceptance in the biomedical area [18]. The first two fundamental stages that are taken in natural synthesis are called bio-reduction and bio-adsorption. Bio-reduction is the procedure by which metallic ions are reduced biochemically into their stabilised structure, and bio-adsorption is the procedure of binding cations onto the surface of a microbe, such as the cell

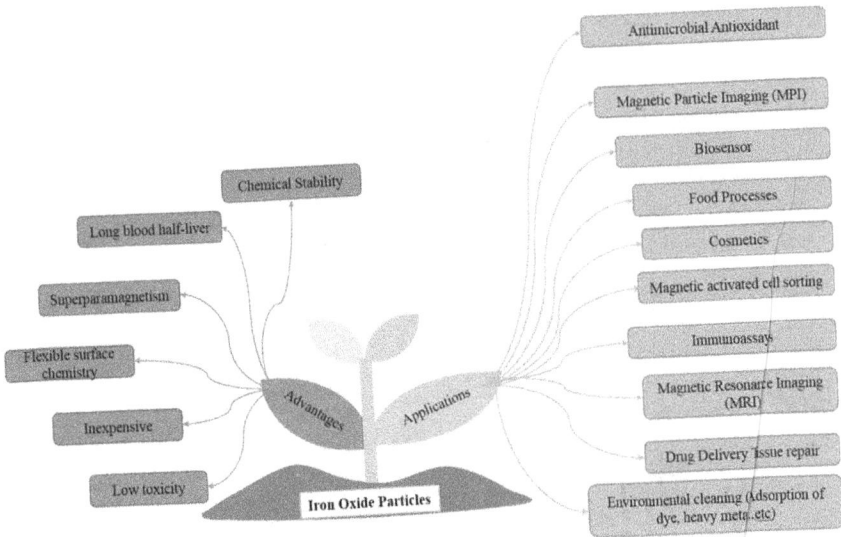

**FIGURE 2.3** Advantages and Versatile Applications of IONPs.

membrane or proteins, to form extra coordination compounds. Both of these processes are referred to collectively as the bio-metalurgical cycle [19].

In addition, the bio-based synthesis does not include any ancillary stages, such as encapsulating or attaching of the molecule with biological activity on their surface, which are required for the production of persistent particles. In conjunction to this, the time required for these types of synthesis procedures is far less than that required for physico-chemical approaches [20]. In this overview, the primary focus is placed on the natural production of IONPs by bacteria, fungi, algae, and plant species (Figure 2.4), as well as an analysis of the benefits and drawbacks associated with each method.

## 2.2 VARIOUS GREEN TECHNOLOGY FOR SYNTHESIS OF IRON OXIDE NANOPARTICLES

### 2.2.1 GREEN SYNTHESIS OF IONPs USING FUNGAL BIOMASS

Fungal extracellular production of iron oxide nanoparticles species is regarded as beneficial when the ease of reproduction is taken into consideration in terms of scale, utilisation of cost-effective raw materials for expansion, high capability for the formation of biomass, ease of down streaming processes, and the toxicity of the residue, as well as economic feasibility [21].

In conjunction with this, species of fungi demonstrate exceptional characteristics such as the attribute of tolerance as well as biomagnification, which helps with the synthesis referring to the nanoparticles of metal [22]. The various interactions between metallic ions and microorganisms have been thoroughly investigated and the ability of microorganisms to do so has led them to be utilised to extract heavy metals from a variety of sources. It is well-known that fungi can produce a significant

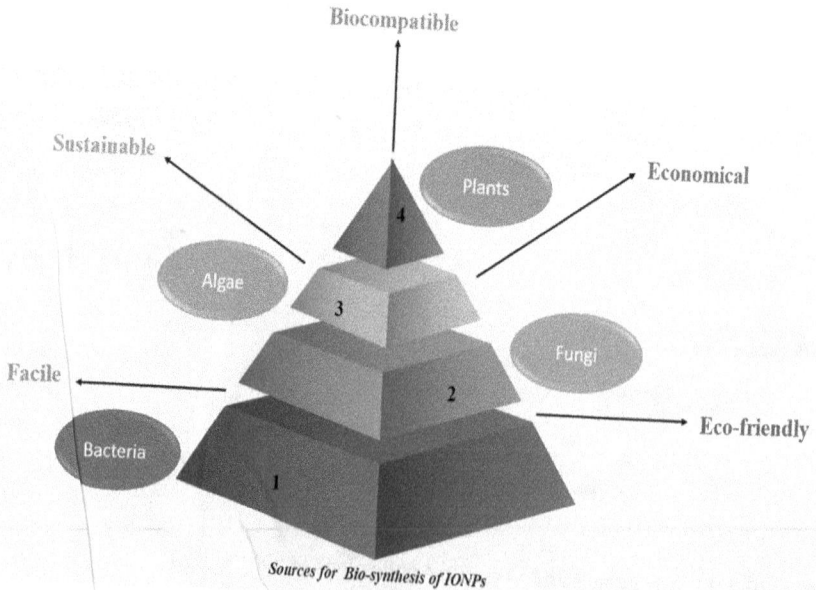

**FIGURE 2.4** Significance of Biosynthesis of IONPs from Different Sources.

quantity of enzymes outside of the cels, which is capable of hydrolyzing a variety of metals.

It is hypothesised that the production of wide variety of extra-cellularly produced nanoparticles is due to the enzyme-based duction of metallic ions. According to the researchers, nitrate dehydrogenase is an example of this type of enzyme that can be found in fungi [23]. Despite this, the damental mechanisms involved in the manufacture of metal nanoparticles (MNPs) tilising fungus are not yet totally understood. The biological material that is employ and the process conditions that are applied are thought to be essential components the production of MNPs by fungi. The bio-based material consists of the strain of robe, the growth variables, and the technique of specimen preparation. On the other d, the process conditions place an emphasis on the metal catalyst that was utilised, th ensity of the precursor solution, temperature, and the pH of the reaction [24].

In the course of researching the topic, one would discover th ere is a significant amount of variation concerning the starting bio-material that is a ied in biosynthe-sis. For the synthesis of the nanoparticles, some methods have m use of fungal biomass, while others have made use of fungal homogenate or the use of fungal residue of the culture. The procedure that was used the most frequ free filtered cultivating the fungus in a growing medium that was suitable for its ne required centrifuging the fungal mycelia that had been produced. After obtaining t and then in this manner, they are combined with a metallic precursors salt and left to ycelia at 28°C for four or five days. Additional filtration of this solution results in t ate mation of a suspension that is composed of manufactured NPs [25].

Pure iron and a substantial fraction of spherical nanoparticles composed of the element $Fe_3O_4$ were produced when homogenate mycelia of *Aspergillus niger* was used to synthesise nanoparticles in the presence of supercritical ethanol. According to the results of the FESEM investigation, the average size of iron nanoparticles is 18 nm, whereas the average size of $Fe_3O_4$ nanoparticles is 50 nm. The magnetic property measurements demonstrate that Fe and $Fe_3O_4$ nanoparticles exhibit super-paramagnetic and ferro-magnetic-like behaviour [24].

## 2.2.2    GREEN SYNTHESIS OF IONPS USING ALGAL BIOMASS

Algae are characterised as photosynthetic organisms without a root or leaf structure. Microalgae, which are unicellular and macro-algae, also known as seaweeds, are multicellular and are both frequently used in the domain of nano-science for the production of various types of nano particles, including copper, gold, silver, iron, and palladium [26]. Similar to plant species and bacteria, algal blooms are enterprises that produce a wide variety of biomolecules such as polypeptides, fats, complex carbs, proteins, alkaloids, macrolides, triterpenoids, cell membrane polysaccharides, polypeptides (contained different functional groups such as hydroxyl-OH, carbonyl –C=O and carboxyl–COOH). The enzymes play a crucial role in reducing, blocking, fabricating, and stabilising the NPs. As a result, this strategy for producing nanoparticles is regarded as one that is risk-free, straightforward, financially advantageous, and pollution free.

In the same manner, while dealing with algal biomass, the seaweeds are first cleaned using distilled water, and then they are dried [26]. In contrast, the bleaching stage in the protocol developed comes right after the wash cycle, and the specimen is therefore exposed to the sun for three days [27]. The specimens are then ground, freeze-dried [28], quantified, combined with the necessary amount of distilled water, and heated [29]. Interpreting the moiety, crystalline structure, thickness, morphology, and magnetic features of produced nanoparticles can be accomplished with the assistance of all of aforementioned techniques.

## 2.2.3    GREEN SYNTHESIS OF IONPS USING BACTERIA

As a significant model in the field of nano science, the bacterial system has also been the subject of extensive research due to its pervasive existence, rapid multiplication time, ability to proliferate under harsh conditions, as well as cost-effective and simple substrates for culture [30]. This technique has been regarded as the most effective method for the synthesis of nanoparticles with a wide variety of forms, volume fraction, and structural frameworks, as well as chemical and physical properties, because the action of the reductase enzymes in microbes may both accumulate and de-toxicate metallic ions. This ability has been utilised for the purpose of producing metallic nanoparticles by employing salts as reaction substrates [31].

In the course of the synthesis method, the reduction of metallic ions is an essential step. It is controlled by a wide variety of parameters, including the presence of ligands on the cell membrane (which is necessary for bio-mineralisation), the kind of strains, environmental factors such as pH and temperature, the growing medium, the proportion of salts, and so on. The dimensions, morphology, content, and output of nanoparticles are all affected, either indirectly or directly by all of the aspects that have been mentioned [32]. Because this environmentally friendly method of synthesis is controlled by an enzyme-mediated system for the reduction of metallic ions, the heat, pH, and culturing period for the bacterial consortium are the most important issues to consider. Biosynthesis is dependent on the presence of optimal conditions. The enzymes are rendered fully inactive and their activity is halted when the pH and temperature are increased [33]. This results in a decrease in the rate at which nanoparticles are formed. In case of Gram-positive bacteria, the cellular wall is made up of glycoprotein, enzymes, carbohydrates, and lipoteichoic and teichoic acid, all of which are capable of acting as a receptor for the bio-adsorption and bio-reduction of metallic ions.

### 2.2.4   GREEN SYNTHESIS OF IONPS USING PLANT SPECIES

For the production of a wide variety of nanostructures, plant species are typically regarded as an excellent source of material due to their low cost, ease of handling, lack of potential for harm, and availability [34]. The biosynthesis procedure utilised distinct parts such as root system, seeds, leaves, flowers, fruit and vegetables, peels, carpels, the entire plant, and grain husk because these are enriched with various biological molecules such as carbs, peptides, polyphenols, proteins, tannins, saponins, terpenes, and nutrients that perform as reducers, stabilisers, redox facilitators, and encapsulating agents in the biosynthetic pathways of NPs [35]. Nanoparticles (NPs) that are synthesised through using plant extracts are speculated to be significantly more stable than NPs that are traditionally generated [36]. NPs that are produced using plant extracts are also typically of distinguishable shapes, such as rounded, cubical, tubular, needle-like, stem-like, pyramidal, and dendritic [37]. Figure 2.4 displays the significance of biosynthesis protocols of IONPs from different bio-based sources.

The characteristics of synthesised nanoparticles are largely determined by discrete criteria such as the kind of plant extract used, the volume fraction of the extract to the metallic precursors medium, and the process conditions (pH, temperature, and length of experimental period) [38]. Therefore, on the grounds of these outlined variables, various methods of biosynthesis have been utilised by researchers. Figure 2.5 shows schematic illustration for biosynthesis protocol of IONPs from biological sources.

Such methods differ slightly from each other in terms of selection of the starting material, technique for producing plant extract, choice of iron salt, existence of NaOH, varying reaction conditions, and the technique of collecting the synthesised NPs. The simplified protocols for IONPs biosynthesis are shown in Figure 2.6.

## 2.3   APPLICATIONS OF IRON OXIDE NANOPARTICLES FOR WATER TREATMENT

### 2.3.1   IRON OXIDE NANOPARTICLES AS ADSORBENT FOR HEAVY METALS

Due to their toxicity to plants, animals, and humans and their propensity to bio-accumulate even at comparatively low concentrations, heavy metal poisoning is a major

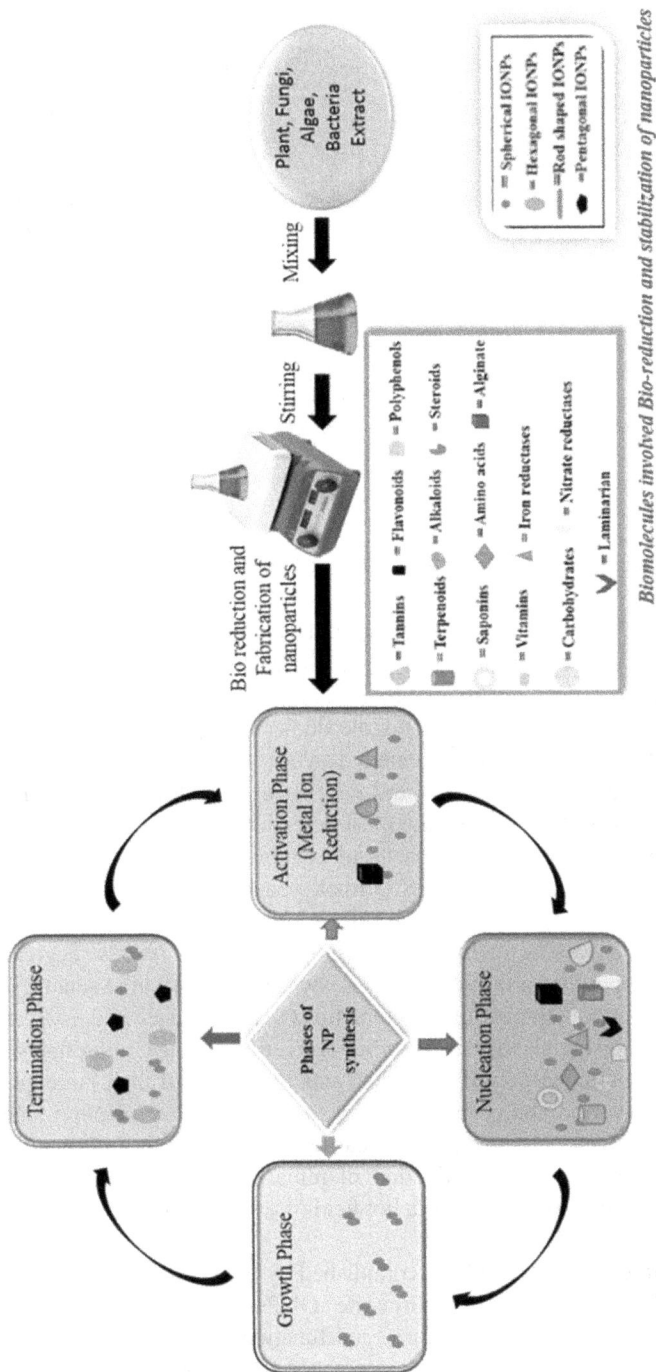

**FIGURE 2.5**    Schematic Illustration for Biosynthesis Protocol of IONPs from Biological Sources.

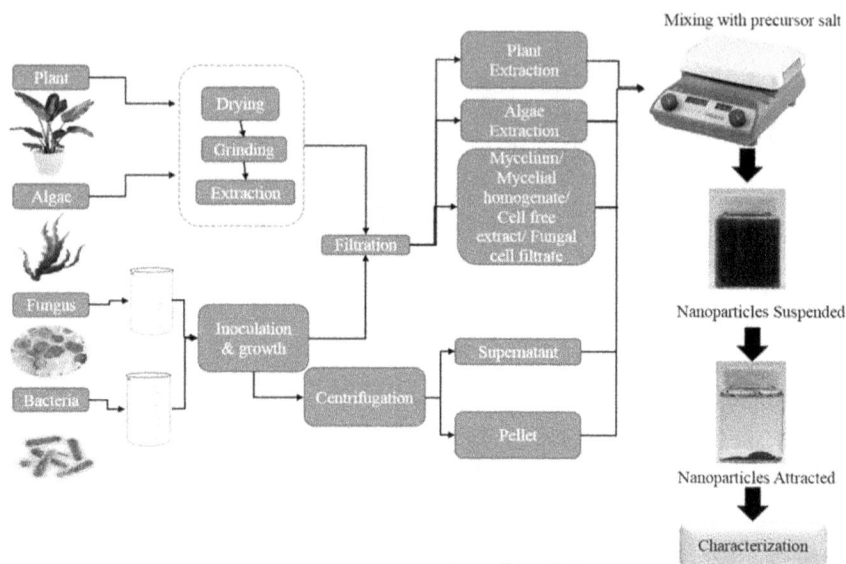

**FIGURE 2.6**   Simplified Protocol for Synthesis of IONPs Using Green Routes.

reason for concern. Therefore, it is of the utmost importance to develop efficient elimination methods for heavy metal ions, which have drawn a significant amount of interest in both theoretical and applied research [39].

At this time, the overwhelming bench-scale studies and applications of nanomaterials for wastewater treatment have concentrated on magnetic nanomaterials [40], carbon nanotubes [41], activated carbon [42], and zero-valent iron [43]. Iron oxide magnetic nanoparticles (IONPs) appear to be the most potential candidate for the remediation of heavy metals, as they have the capacity to treat huge volumes of wastewater and are easy to use for magnetic isolation. Nevertheless, agglomeration generated by large surface active area to volume ratios of IONPs has the potential to affect a variety of crucial ecological processes, particularly metal uptake. This is because aggregation is one of the most important surface-driven phenomena that can occur in aquatic settings.

In addition to agglomeration, other interactions that take place in wastewater also have an effect on the metals that are adsorbed there. For instance, phosphates ($PO_4^{3}$) have the potential to be effectively absorbed and, as a result of their large quantities in wastewater, have the ability to overpower metallic ions for sorption sites [44]. Consequently, the factors that were noted earlier may be able to limit the effectiveness of nano-sorbents. The investigation of remarkably effective reconfiguration methodologies for IONPs tends to be a hot field of research for increasing the effectiveness of nano-sorbents.

Surface treatment, which can be accomplished by the bonding of synthetic shells and/or bio molecules, not only stabilises the IONPs but also eventually precludes its oxidation, delivers particular features and functionality that can be specific for ion take-up, and, as a result, enhances the potency for heavy metal capture in water treatment practises.

## 2.3.2 IRON OXIDE NANOPARTICLES AS ADSORBENT FOR ORGANIC CONTAMINANTS

Adsorption is a well-established separation technology that has seen extensive use in the industry for the purpose of removing chemical contaminants from water. It has a number of advantages in terms of affordability, adaptability, and simplicity of development and maintenance, as well as lack of sensitivity to harmful contaminants [45]. For the removal of organic contaminants, it is desired to have access to a sorbent that is both efficient and cost-effective and that also has a large capacity for adsorption. Iron oxide nanoparticles (IONPs) are now being investigated for their potential use in the sorption of organic contaminants, particularly for the effective remediation of large volume of water samples and the speedy separation that may be achieved by utilising a powerful external magnetic field. Numerous studies [46] have been conducted to investigate the effectiveness of iron oxide nanomaterials (IONPs) for the elimination of organic pollutants. For instance, it was discovered that $Fe_3O_4$ hollow nano-spheres are an efficient sorbent for red dye (with a maximum adsorption capacity of 90 mg $g^{-1}$). It was found that the manufactured nano-spheres had a saturation magnetisation of 42 emu $g^{-1}$, which was adequate for magnetic isolation [47].

In conclusion, the coupling of the outstanding adsorption capability and magnetic characteristics of IONPs is generally considered to be a promising strategy to cope with a multitude of environmental issues. The development of new materials containing iron oxide could open the door to the development of adsorption systems of the next generation that have a high capacity, are simple to separate, and have longer lifecycles. Overall adsorption and photocatalytic pollutant degradation pathways by using IONPs are illustrated by Figure 2.7.

## 2.3.3 IRON OXIDE NANOPARTICLES AS PHOTO-CATALYST

The photocatalytic activity involves introducing UV/visible irradiance and the relevant nanostructures on the polluted area [48]. It is another promising pathway for the elimination of pollutants and contaminants from wastewater. In this process, the pollutants can be progressively oxidised into lower molecular weight transitional substances and subsequently converted into $CO_2$, $H_2O$, and anions ($NO_3$, $Cl^{-1}$ and $PO_4^3$) [49]. It has been demonstrated that the photo-Fenton process is both an efficient and a favourable research for the disinfection of microbes that are present in wastewater. Nevertheless, for optimal results, solutions with pH values close to neutral are suggested rather than using extremely basic or acidic solution [50].

It has been suggested that a number of different species of Fe(III) oxides, such as α-$Fe_2O_3$, β-$Fe_2O_3$, α-FeOOH, β-FeOOH, and γ-FeOOH, might break down organic contaminants and lessen their cytotoxicity as a result of an improved photocatalytic activity effect [51]. These oxides include α-$Fe_2O_3$ and γ-$Fe_2O_3$. These NPs are an example of a revolutionary method for manipulating the catalytic features of IONPs for photocatalysis, which is a step towards developing a nano enhanced technology that is both harmless and efficient for treating wastewater. An instance of this was the photo degradation of Congo red (CR) by IONPs [52]. When the particles were 100 nm in size, the removal effectiveness was at its highest, 96%.

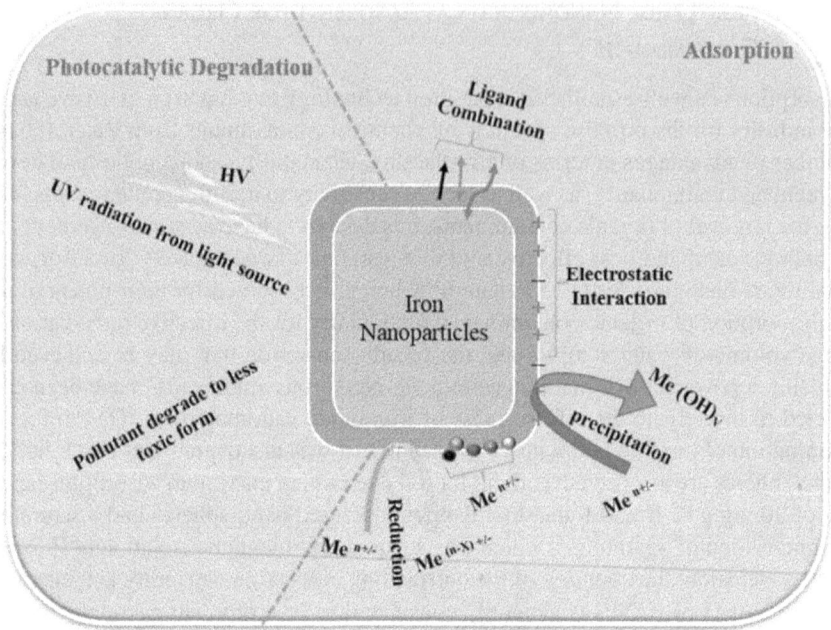

**FIGURE 2.7**   Adsorption and Photocatalytic Degradation of Pollutants Using IONPs.

Irradiation of the $Fe_2O_3$ nanoparticle results in the production of a pair consisting of an electron ($e^-$) and a hole (h+). The electron is then excited from the valence band towards the conduction band, while the hole remains in the valence band (VB). The hole, shown by the symbol h+, is accountable for the transformation of $H_2O$ into the hydroxyl radical (–OH), which in turn is accountable for the oxidative destruction of the dye. On the other hand, unbound electrons react with molecular oxygen to produce superoxide radicals when they are exposed to oxygen molecules. Hydroxyl radicals are produced via the transformation of the superoxide radical. The hydroxyl radical is a powerful oxidising agent that destroys organic species into nontoxic end products in a non-selective manner (Figure 2.8) [53].

## 2.4   CHALLENGES AND FUTURE PERSPECTIVE OF USING IRON OXIDE NANOPARTICLES

It indicates that metal-oxide based nanoparticles (MO) will play a major role in wastewater remediation applications in the twenty-first century. The nanoparticles can indeed be generated by a number of different synthetic and bio-based techniques. Nevertheless, in order to bring down prices, it will be necessary to manufacture these nanoparticles, including iron oxide nanoparticles (IONPs), on a commercial level for mass production. The natural resources used in the production of the MO nanoparticles should really be able to be maintained for an extended period of time, have cheap cost, be harmless to the environment, and not include any hazardous substances. Producing nanoparticles with a single distribution is essential for the

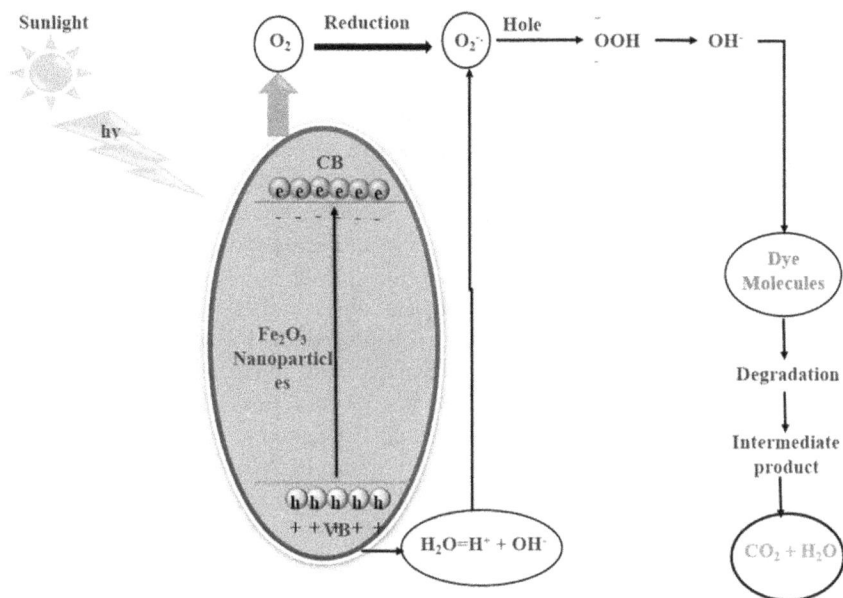

**FIGURE 2.8**   Photocatalytic Degradation Pathways of Dye Molecules Using IONPs.

continuation of scientific investigation. Nevertheless, the mechanism underlying the synthesis of these MO nanoparticles, including IONPs, is not well understood at this time. Therefore, the process by which MO nanoparticles can be manipulated in terms of their size and form should be the primary focus of research in the future. Another significant obstacle is the requirement to lessen the hazardous effects of nanoparticles while simultaneously increasing their use for the environmental applications. Better methods are being proposed to address such issues by bio synthesis of metal nanoparticles through breakthroughs in nanotechnologies. Nevertheless, before their widespread deployment, their effects on biological health parameters need to be carefully considered in light of the fact that the majority of nano-materials, until now, have been fairly affordable in comparison to conventional substances such as commercial activated carbon. The primary focus of potential developments will be on optimising the performance of various processes, which will require just trace amounts of nanomaterials made of IONPs. Furthermore, additional research needs to be conducted in order to devise techniques of synthesis that are efficient in terms of expense, and extensive testing is essential for the practical ground application of IONPs nanoparticles.

The fundamental focus of research in the domain of nano-science is for the production of biocompatible nano scale particles. The presence of a broad range of biological molecules, demonstrating significant reduction of metallic ions to produce nanoparticles from the metal matrix, makes green sources for the bio-mediated synthesis of IONPs. This is mainly owing to the fact that green credible sources are indeed the major players in the process. There are a variety of factors that contribute to the economically lucrative nature of biological nanoparticle synthesis. To begin, this turns out to be a very cost-effective method due to the fact that the synthesis just needs basic physical conditions, does not require any complex instruments, and makes use of biological agents that are readily available. The lack of hazardous substances that are mandatory for the production of those particles is the next important factor to take into consideration, as it is what makes them biocompatible and safe for the environment. Moreover, water is the primary solvent utilised in the majority of the plant-mediated strategies, which adds additional significance to the term "green technology." Because the biomolecules that are present in the extract are capable of carrying out this action on their own, there is no aggregation observed. This means that no other distinct capping entities are needed for the stabilisation of the biochemically synthesised NPs. However, the greatest obstacle that this methodology must overcome is the difficulty of reproducing this procedure. This method takes a static technique towards the formation of the nanoparticles (NPs) by using the reactants partially. The un-reacted ingredients have the potential to result in the development of unexpected compounds with a wide variety of features, which would make the procedure impossible to reproduce. The fast retrieval of the products has the potential to address the issue to a limited extent; however, it continues to be a challenging factor for the research. Researchers are mandated to give attention to the production of NPs with desired size because it is the central aspect that determines the attributes of NPs and, subsequently, their applicability in versatile field. In addition, a compilation of the most recent research demonstrates that innumerable questions regarding the variables that control the dimensions, structure, crystalline nature, output of the NPs obtained, and accurate magnetic characteristics of the NPs manufactured utilising a green, sustainable protocol are not yet answered. Therefore, additional research in this domain is absolutely important to bridge these gaps and provide more randomised trials.

There has not been a lot of in-depth research done on how NPs affect the human health or the environment. Some NPs have detrimental effects because of their size or special chemical properties, especially the ones which are freely transportable and are not integrated into a medium [54, 55]. These NPs are responsible for the majority of the negative effects. The increasing usage of nanomaterials based on iron will undoubtedly result in the release of trace quantities into the water habitats, which will, in turn, induce contact with live species and may result in health concerns [56]. Iron oxide nanoparticles (IONPs) are efficient nanomaterials for the removal of numerous contaminants from wastewater. Nevertheless, utilising nano-sized IONPs as an adsorbent comes with the immediate disadvantage as the flow of water in continuous adsorption columns over the filter bed requires a pressure head. Methodologies using open columns, in which nanoparticles are encased inside a matrix of organic or inorganic polymers and beads, need to be developed. It is also possible for the substrate matrix to change the structural features of the NPs, which means that these changes need to be taken advantage of in order to remove contaminants from wastewater [57]. In addition, recent studies have talked about the possibility of iron-based nanoparticles being cytotoxic, along with the possibility of their having repercussions in the fields of medicine, chemistry, and physiology [58].

To this day, the most significant challenges that researchers face when trying to develop NPs from a prototype or lab scale process are the regeneration and recycling of NPs up to the commercialisation level [59]. The most significant drawback associated with all these bio-based processes is the possibility that the chemical make-up of the extracts might be modified in the biomass resources, even among the same species, as a result of differences in the environmental conditions. There are a number of factors that can limit their uses, including – but not limited to – seasonal change, differences in geography, and different growth stages of each other. This, in turn, can impede the development of photo-green approaches. However, this drawback can be avoided by first detecting and then separating the constituents that are present inside the bio extracts and afterwards employing those components again for goal of water purification. As a result, investigations need to be carried out in this particular direction.

Earlier research findings illustrated the potential drawbacks of utilising zero valent iron (ZVI) nanoparticles (IONPs) for the purification of wastewater [60]. According to what the researchers reported, there are a few issues that are connected to the procedure that must be resolved. For instance, fast particle agglomeration took place as a response to applying magnetic pull to IONPs. In addition to this, the ZVI nanoparticle has a higher propensity to interact with air and molecules that include oxygen. In the meantime, ZVI nanoparticles might be hazardous to some microbial species, and researchers are getting closer to understanding how this might play out at the subcellular or community level.

The parting of iron ox-hydroxide (FeOOH) nanoparticles from the last solution after the reaction is yet an additional endeavour looking for the inherent problem of utilising it as a sorbent. The finely grounded colloidal solution is formed from the suspension of FeOOH which makes it difficult and expensive to pile up FeOOH NPs from the ultimate wastewater effluents [61]. These are the primary potential downsides of nanoscale zero-valent IONPs (ZVI). Other disadvantages include the discharge of dissolved iron ions and its genetic susceptibility towards the environment.

In order to circumvent these challenges, the ZVI NPs were stabilised with the use of appropriate solid substrates, and the parameters that influence the Fenton reaction, most notably pH, were adjusted. They also outline difficulties that might arise as a consequence of variations in the physico-chemical characteristics of ZVI NPs, due to their amendment. Research on toxicity reveals that cellular membrane rupture and oxidative stress caused by the formation of oxygen-enriched species and $Fe^{2+}$ by ZVI nanoparticles are the primary contributors to the cytotoxicity [62].

## 2.5 CONCLUSION

Within the scope of this chapter, a variety of environmentally friendly strategies for producing IONPs from a wide range of plants and microbes was reviewed. In light of the comparison between these two methods of synthesis, the approach including the plant extract comes out on top. This could be a reference to its ease of use, its short reaction time, and its tolerability in comparison to techniques of synthesis relying on microorganisms, which could result in infections or products with a high level of toxicity. Nanoparticles of various forms and morphologies, which had been used in a wide variety of applications, were produced by environmentally friendly synthesis processes. The findings of this chapter indicate that the environmentally friendly approaches are more stable than many of the chemical procedures. Recently, green synthesis protocols have gained popularity due to their simplicity, safety, low cost, and positive impact on the environment. Iron nanoparticles were manufactured and then employed in a variety of applications. Some of the IONPs were used in bioremediation, while others were engaged because of their antibacterial function. The use of these technologies will be of tremendous assistance in preventing the polluting of our natural resources.

This findings show that a variety of IONPs-based technologies have been proposed or are now being actively developed for the treatment of wastewater; however, the majority of these technologies are still in the trial or prototype stage of development. There is a possibility that complications will arise during the use of in in vivo and in vitro research with IONPs. The study of IONPs in a wide variety of structural and chemical configurations has previously demonstrated its versatility and the possible applications for elimination of emerging pollutants from aqueous effluents.

It should be highlighted, however, that before all these nanomaterials may be used on a mass scale, there are many unanswered questions regarding their effects on human health and the ecosystem that need to be answered. Because discharges are already happening to the environment, it is becoming increasingly vital to understand their course of events and the influence that they have on the surrounding environment. Expanding research in this field is necessary since there is a high probability that the amount of IONPs discharged will continue to rise, along with the rapid expansion of the sector and the enormous information gaps that exist in hazards evaluation and management. To summarise, there has been a lot of renewed interest in the application of designed IONPs as a fairly non-invasive, in-situ technique for applications in the treatment of wastewater.

## ACKNOWLEDGMENTS

The authors are thankful for the funding provided by Malaysian Joint Research Scheme-ST 077–2022, Interdisciplinary Research IIRG003A-2022IISS and International Grant ICF 023–2022 and ICF 080–2021 under University of Malaya, Kuala Lumpur 50603, Malaysia.

## REFERENCES

[1]  Arya, A., Mishra, V., and Chundawat, T. S. 2019. Green Synthesis of Silver Nanoparticles from Green Algae (*Botryococcus Braunii*) and its Catalytic Behaviour for the Synthesis of Benzimidazoles. *Chemical Data Collections*, 20, 100190. doi:10.1016/j.cdc.2019.100190.

[2]  Hussain, M., Raja, N. I., Iqbal, M., and Aslam, S. 2019. Applications of Plant Flavonoids in the Green Synthesis of Colloidal Silver Nanoparticles and Impacts on Human Health. *Iranian Journal of Science and Technology, Transaction A, Science*, 43 (3), 1381–1392. doi:10.1007/s40995-017-0431-6.

[3]  Mourdikoudis, S., Pallares, R. M., and Thanh, N. T. 2018. Characterization Techniques for Nanoparticles: Comparison and Complementarity Upon Studying Nanoparticle Properties. *Nanoscale*, 10 (27), 12871–12934. doi:10.1039/C8NR02278J.

[4]  Thakkar, K. N., Mhatre, S. S., and Parikh, R. Y. 2010. Biological Synthesis of Metallic Nanoparticles. *Nanomedicine: Nanotechnology, Biology and Medicine*, 6 (2), 257–262. doi:10.1016/j.nano.2009.07.002.

[5] Khan, A. U., Khan, M., Malik, N., Cho, M. H., and Khan, M. M. 2019. Recent Progress of Algae and Blue–green Algae-Assisted Synthesis of Gold Nanoparticles for Various Applications. *Bioprocess Biosystem Engineering*, 42 (1), 1–15. doi:10.1007/s00449-018-2012-2.

[6] Revati, K., and Pandey, B. D. 2011. Microbial Synthesis of Iron-Based Nanomaterials—A Review. *Bulletin of Materials Science*, 34 (2), 191–198. doi:10.1007/s12034-011-0076-6.

[7] Aminabad, N. S., Farshbaf, M., and Akbarzadeh, A. 2019. Recent Advances of Gold Nanoparticles in Biomedical Applications: State of the Art. *Cell Biochemistry and Biophysics*, 77 (2), 123–137. doi:10.1007/s12013-018-0863-4.

[8] Ali, A., Zafar, H., Zia, M., ul Haq, I., Phull, A. R., Ali, J. S., and Hussain, A. 2016. Synthesis, Characterization, Applications, and Challenges of Iron Oxide Nanoparticles. *Nanotechnology Science Application*, 9, 49–67. https://doi.org/10.2147/NSA.S99986.

[9] Sruthi, P. D., Sahithya, C. S., Justin, C., SaiPriya, C., Bhavya, K. S., Senthilkumar, P., et al. 2019. Utilization of Chemically Synthesized Super Paramagnetic Iron Oxide Nanoparticles in Drug Delivery, Imaging and Heavy Metal Removal. *Journal of Cluster Science*, 30 (1), 11–24. doi:10.1007/s10876-018-1454-7.

[10] Bhuiyan, M. S. H., Miah, M. Y., Paul, S. C., Aka, T. D., Saha, O., Rahaman, M. M., et al. 2020. Green Synthesis of Iron Oxide Nanoparticle Using Carica Papaya Leaf Extract: Application for Photocatalytic Degradation of Remazol Yellow RR Dye and Antibacterial Activity. *Heliyon*, 6 (8), e04603. doi:10.1016/j.heliyon. 2020. e04603.

[11] Rauwel, P., Küünal, S., Ferdov, S., and Rauwel, E. 2015. A Review on the Green Synthesis of Silver Nanoparticles and Their Morphologies Studied via TEM. *Advanced Materials Science and Engineering*, 2015, 1–9. doi:10.1155/2015/682749.

[12] Gao, L., Fan, K., and Yan, X. 2020. Iron Oxide Nanozyme: A Multifunctional Enzyme Mimetics for Biomedical Application. *Nanozymology*, 105–140. doi:10.1007/978-981-15-1490-6_5.

[13] Jubran, A. S., Al-Zamely, O. M., and Al-Ammar, M. H. 2020. A Study of Iron Oxide Nanoparticles Synthesis by Using Bacteria. *International Journal of Pharmaceutical Quality Assurance*, 11 (01), 01–08.

[14] Wu, W., He, Q., and Jiang, C. 2008. Magnetic Iron Oxide Nanoparticles: Synthesis and Surface Functionalization Strategies. *Nanoscale Research Letters*, 3 (11), 397. doi:10.1007/s11671-008-9174-9.

[15] Salam, H. A., Rajiv, P., Kamaraj, M., Jagadeeswaran, P., Gunalan, S., and Sivaraj, R. 2012. Plants: green Route for Nanoparticle Synthesis. *International Research Journal of Biological Science*, 1 (5), 85–90.

[16] Latif, M. S., Abbas, S., Kormin, F., and Mustafa, M. K. 2019. Green Synthesis of Plant-Mediated Metal Nanoparticles: The Role of Polyphenols. *Asian Journal of Pharmaceutical and Clinical Research*, 12 (7), 75–84. doi:10.22159/ajpcr.2019.v12i7.33211.

[17] Vasantharaj, S., Sathiyavimal, S., Senthilkumar, P., LewisOscar, F., and Pugazhendhi, A. 2019. Biosynthesis of Iron Oxide Nanoparticles Using Leaf Extract of Ruellia Tuberosa: Antimicrobial Properties and Their Applications in Photocatalytic Degradation. *Journal of Photochemistry Photobiology B: Biology*, 192, 74–82. doi:10.1016/j.jphotobiol.2018.12.025.

[18] Gholampoor, N., Emtiazi, G., and Emami, Z. 2015. The Influence of Microbacterium Hominis and Bacillus Licheniformis Extracellular Polymers on Silver and Iron Oxide Nanoparticles Production; Green Biosynthesis and Mechanism of Bacterial Nano Production. *Journal of Nanomaterials & Molecular Nanotechnology*, 04 (02). doi:10.4172/2324-8777.1000160.

[19] Pantidos, N., and Horsfall, L. E. 2014. Biological Synthesis of Metallic Nanoparticles by Bacteria, Fungi and Plants. *Journal of Nanomedicine and Nanotechnology*, 5 (5), 1. doi:10.4172/2157-7439.1000233.

[20] Gahlawat, G., and Choudhury, A. R. 2019. A Review on the Biosynthesis of Metal and Metal Salt Nanoparticles by Microbes. *RSC Advances*, 9 (23), 12944–12967. doi:10.1039/C8RA10483B.

[21] Tarafdar, J. C., and Raliya, R. 2013. Rapid, Low-Cost, and Ecofriendly Approach for Iron Nanoparticle Synthesis Using Aspergillus Oryzae TFR9. *Journal of Nanoparticles*, 2013, 1–4. doi:10.1155/2013/141274.

[22] Agarwal, H., Kumar, S. V., and Rajeshkumar, S. 2017. A Review on Green Synthesis of Zinc Oxide Nanoparticles–An Eco-Friendly Approach. *Resource- Efficient Technology*, 3 (4), 406–413. doi:10.1016/j.reffit.2017.03.002.

[23] Abdeen, M., Sabry, S., Ghozlan, H., El-Gendy, A. A., and Carpenter, E. E. 2016. Microbial-physical Synthesis of Fe and $Fe_3O_4$ Magnetic Nanoparticles Using Aspergillus Niger YESM1 and Supercritical Condition of Ethanol. *Journal of Nanomaterials*, 2016, 1–7. doi:10.1155/2016/9174891.

[24] Silva, L. P., Reis, I. G., and Bonatto, C. C. 2015. Green Synthesis of Metal Nanoparticles by Plants: Current Trends and Challenges. *Green Processing Nanotechnology*, 259–275. doi:10.1007/978-3-319-15461-9_9.

[25] Bhargava, A., Jain, N., Barathi, M., Akhtar, M. S., Yun, Y. S., and Panwar, J. 2013. Synthesis, Characterization and Mechanistic Insights of Mycogenic Iron Oxide Nanoparticles. *Nanotechnology Sustainable Development*, 15, 337–348. doi:10.1007/978-3- 319–05041–6_27

[26] Pang, Y., Zeng, G. M., Tang, L., Zhang, Y., Liu, Y. Y., Lei, X. X., et al. 2011. Cr(VI) Reduction by Pseudomonas Aeruginosa Immobilized in a Polyvinyl Alcohol/Sodium Alginate Matrix Containing Multi-Walled Carbon Nanotubes. *Bioresource Technology*, 102, 10733–1076.

[27] Yew, Y. P., Shameli, K., Miyake, M., Kuwano, N., Khairudin, N. B. B. A., Mohamad, S. E. B., et al. 2016. Green Synthesis of Magnetite ($Fe_3O_4$) Nanoparticles Using Seaweed (Kappaphycus Alvarezii) Extract. *Nanoscale Research Letters*, 11 (1), 1–7. doi:10.1186/s11671-016-1498-2.

[28] Mahdavi, M., Namvar, F., Ahmad, M. B., and Mohamad, R. 2013. Green Biosynthesis and Characterization of Magnetic Iron Oxide ($Fe_3O_4$) Nanoparticles Using Seaweed (Sargassum Muticum) Aqueous Extract. *Molecules*, 18 (5), 5954–5964. doi:10.3390/molecules18055954

[29] Siji, S., Njana, J., Amrita, P. J., and Vishnudasan, D. 2018. Biogenic Synthesis of Iron OxideNanoparticles from Marine Algae. TKMInt. *Journal of Multidisciplinary Research*, 1 (1), 1–7.

[30] Fariq, A., Khan, T., and Yasmin, A. 2017. Microbial Synthesis of Nanoparticles and Their Potential Applications in Biomedicine. *Journal of Applied Biomedicine*, 15 (4), 241–248. doi:10.1016/j.jab.2017.03.004.

[31] Alam, H., Khatoon, N., Khan, M. A., Husain, S. A., Saravanan, M., and Sardar, M. 2020. Synthesis of Selenium Nanoparticles Using Probiotic Bacteria Lactobacillus Acidophilus and Their Enhanced Antimicrobial Activity against Resistant Bacteria. *Journal of Cluster Science*. 31 (5), 1003–1011. doi:10.1007/s10876-019-01705-6.

[32] Iravani, S., and Varma, R. S. 2020. Bacteria in Heavy Metal Remediation and Nanoparticle Biosynthesis. *ACS Sustainable Chemistry and Engineering*, 8 (14), 5395–5409. doi:10.1021/acssuschemeng.0c00292.

[33] Garg, D., Sarkar, A., Chand, P., Bansal, P., Gola, D., Sharma, S., et al. 2020. Synthesis of Silver Nanoparticles Utilizing Various Biological Systems: Mechanisms and Applications—A Review. *Progress Biomaterials*, 9 (3), 1–15. doi:10.1007/s40204-020-00135-2.

[34] Noruzi, M. 2015. Biosynthesis of Gold Nanoparticles Using Plant Extracts. *Bioprocess Biosystem Engineering*, 38 (1), 1–14. doi:10.1007/s00449-014-1251-0.

[35] Rajeshkumar, S., and Bharath, L. V. 2017. Mechanism of Plant-Mediated Synthesis of Silver Nanoparticles–A Review on Biomolecules Involved, Characterisation and Antibacterial Activity. *Chemico-Biological Interactions*, 273, 219–227. doi:10.1016/j.cbi.2017.06.019.

[36] Bibi, I., Nazar, N., Ata, S., Sultan, M., Ali, A., Abbas, A., et al. 2019. Green Synthesis of Iron Oxide Nanoparticles Using Pomegranate Seeds Extract and Photocatalytic Activity Evaluation for the Degradation of Textile Dye. *Journal of Materials Research Technology*, 8 (6), 6115–6124. doi:10.1016/j.jmrt.2019.10.006.

[37] Abdullah, J. A. A., Eddine, L. S., Abderrhmane, B., Alonso-González, M., Guerrero, A., and Romero, A. 2020. Green Synthesis and Characterization of Iron Oxide Nanoparticles by Pheonix Dactylifera Leaf Extract and Evaluation of Their Antioxidant Activity. *Sustainable Chemistry Pharmacology*, 17, 100280. doi:10.1016/j.scp.2020.100280.

[38] Singh, P., Kim, Y. J., Zhang, D., and Yang, D. C. 2016. Biological Synthesis of Nanoparticles from Plants and Microorganisms. *Trends Biotechnology*, 34 (7), 588–599. doi:10.1016/j.tibtech.2016.02.006.

[39] Huang, S. H., and Chen, D. H. 2009. Rapid Removal of Heavy Metal Cations and Anions from Aqueous Solutions by an Amino-Functionalized Magnetic Nano-Adsorbent. *Journal of Hazardous Materials*, 63 (1), 174–179.

[40] Iram, M., Guo, C., Guan, Y. P., Ishfaq, A., and Liu, H. Z. 2010. Adsorption and Magnetic Removal of Neutral Red Dye from Aqueous Solution Using $Fe_3O_4$ Hollow Nanospheres. *Journal of Hazardous Materials*, 181 (1–3), 1039–1050.

[41] Shankar, K., Basham, J. I., Allam, N. K., Varghese, O. K., Mor, G. K., Feng, X. J., et al. 2009. Recent Advances in the Use of $TiO_2$ Nanotube and Nanowire Arrays for Oxidative. *Journal of Physical Chemistry C*, 113, 6327–6359.

[42] Kobya, M., Demirbas, E., Senturk, E., and Ince, M. 2005. Adsorption of Heavy Metal Ions from Aqueous Solutions by Activated Carbon Prepared from Apricot Stone. *Bioresource Technology*, 96 (13), 1518–1521.

[43] Ponder, S. M., Darab, J. G., and Mallouk, T. E. 2000. Remediation of Cr(VI) and Pb(II) Aqueous Solutions Using Supported Nanoscale Zero-Valent Iron. *Environmental Science Technology*, 34, 2564–2569.

[44] Feng, Y., Gong, J. L., Zeng, G. M., Niu, Q. Y., Zhang, H. Y., Niu, C. G., et al. 2010. Adsorption of Cd (II) and Zn (II) from Aqueous Solutions Using Magnetic Hydroxyapatite Nanoparticles as Adsorbents. *Chemical Engineering Journal*, 162 (2), 487–494.

[45] Rafatullah, M., Sulaiman, O., Hashim, R., and Ahmad, A. 2010. Adsorption of Methylene Blue on Low-Cost Adsorbents: A Review. *Journal of Hazardous Materials*, 177 (1–3), 80.

[46] Luo, L. H., Feng, Q. M., Wang, W. Q., and Zhang, B. L. 2011. $Fe_3O_4$/Rectorite Composite: Preparation, Characterization and Absorption Properties from Contaminant Contained in Aqueous Solution. *Advanced Materials Research*, 287592–287598.

[47] MA, Z. Y., Guan, Y. P., Liu, X. Q., and Liu, H. Z. 2005. Preparation and Characterization of Micron\Sized Non\Porous Magnetic Polymer Microspheres with Immobilized Metal Affinity Ligands by Modified Suspension Polymerization. *Journal of Applied Polymer Science*, 96 (6), 2174–2180.

[48] Zhu, M., Wang, Y., Meng, D., Qin, X., and Diao, G. 2012. Hydrothermal Synthesis of Hematite Nanoparticles and Their Electrochemical Properties. *Journal of Physical Chemistry C*, 116, 16276–16285. https://doi.org/10.1021/jp304041m.

[49] Haijiao Lu, H. H., Wang, J., Stoller, M., Wang, T., and Bao, Y. 2017. An Overview of Nanomaterials for Water and Wastewater Treatment. *Advanced Materials Science and Engineering*, 1–12, https://doi.org/10.4018/978-1-5225-2136-5.ch001.

[50] Adeleye, A. S., Conway, J. R., Garner, K., Huang, Y., Su, Y., and Keller, A. A. 2016. Engineered Nanomaterials for Water Treatment and Remediation: Costs, Benefits, and Applicability. *Chemical Engineering Journal*, 286, 640–662. https://doi.org/10.1016/j.cej.2015.10.105.

[51] Tihana Čižmar, Vedran Kojic, Marko Rukavina, Lidija Brkljačić, Krešimir Salamon, Ivana Grčić, Lucija Radetić, and Andreja Gajović. 2020. Hydrothermal Synthesis of FeOOH and $Fe_2O_3$ Modified Self-Organizing Immobilized $TiO_2$ Nanotubes for

Photocatalytic Degradation of 1H-Benzotriazole. *Catalysts*, 10, 1–19. doi:10.3390/catal10121371.

[52]   Khedr, M. H., Abdel Halim, K. S., and Soliman, N. K. 2009. Synthesis and Photocatalytic Activity of Nano-Sized Iron Oxides. *Materials Letters*, 63, 598–601. https://doi.org/10.1016/j.matlet.2008.11.050.

[53]   Jerin, V. M., Remya, R., Thomas, M., and Varkey, J. T. 2019. Investigation on the Removal of Toxic Chromium Ion from Waste Water Using $Fe_2O_3$ Nanoparticles. *Materials Today Proceedings*, 9, 27–31. https://doi.org/10.1016/j.matpr.2019.02.032.

[54]   Van der Bruggen, B., and Nassar, N. N. 2013. The Application of Nanoparticles for Wastewater Remediation. *Applications of Nanomaterials for Water Quality*, 52–65. https://doi.org/10.4155/ebo.13.373.

[55]   Khan, I., Saeed, K., and Khan, I. 2019. Nanoparticles: Properties, Applications and Toxicities. *Arabian Journal of Chemistry*, 12, 908–931. https://doi.org/10.1016/j.arabjc.2017.05.011.

[56]   Liu Yanping, G. P., and Tourbin Mallorie, Lachaize Sébastien. 2014. Nanoparticles in Wastewaters: Hazards, Fate and Remediation. *Powder Technology*, 255, 149–146. doi:10.4028/www.scientific.net/MSF.508.621.

[57]   Biftu, W. K., Ravindhranath, K., and Ramamoorty, M. 2020. New Research Trends in the Processing and Applications of Ironbased Nanoparticles as Adsorbents in Water Remediation Methods. *Nanotechnology Environmental Engineering*, 5, 1–12, https://doi.org/10.1007/s41204-020-00076-y.

[58]   Paunovic, J., Vucevic, D., Radosavljevic, T., Mandić-Rajčević, S. I., and Pantic. 2020. Iron-Based Nanoparticles, and Their Potential Toxicity: Focus on Oxidative Stress and Apoptosis. *Chemico Biological Interaction*, 316, 1–4. https://doi.org/10.1016/j.cbi.2019.108935.

[59]   Goutam, S. P., Saxena, G., Roy, D., Yadav, A. K., and Bharagava, R. N. 2020. Green Synthesis of Nanoparticles and Their Applications in Water and Wastewater Treatment, Bioremediation of Industrial Waste for Environmental Safety Volume I: Industrial Waste and Its Management, 349–379. https://doi.org/10.1007/978-981-13-1891-7_16.

[60]   Zhou, L., Li, R., Zhang, G., Wang, D., Cai, D., and Wu, Z. 2018. Zero Valent Iron Nanoparticles Supported by Functionalized Waste Rock Wool for Efficient Removal of Hexavalent Chromium. *Chemical Engineering Journal*, 339, 85–96. https://doi.org/10.1016/j.cej.2018.01.132.

[61]   Kuang, L., Liu, Y., Fu, D., and Zhao, Y. 2017. FeOOH-Graphene Oxide Nanocomposites for Fluoride Removal from Water: Acetate Mediated Nano FeOOH Growth and Adsorption Mechanism. *Journal of Colloid Interface Science*, 490, 259–269. https://doi.org/10.1016/j.jcis.2016.11.071.

[62]   Stefaniuk, M., Oleszczuk, P., and OK, Y. S. 2016. Review on Nano Zerovalent Iron (nZVI): From Synthesis to Environmental Applications. *Chemical Engineering Journal*, 287 https://doi.org/10.1016/j.cej.2015.11.046.

# 3 TiO$_2$ Doped Lignocellulosic Biopolymer for Water Treatment

## Waste to Wealth – A Pathway Towards Circular Economy

*Abu Nasser Faisal, Zaira Zaman Chowdhury, Masud Rana, Ahmed Elsayid Ali, Rahman Faizur Rafique and Rafie Bin Johan*

### 3.1 INTRODUCTION

The purity of the water on the planet will be the most significant obstacle that humanity will endure in the twenty-first century. Despite the fact that water covers approximately 70% of the planet's surface, just 2.5% of it is accessible for commercial, occupational, and household usage [1–3]. Indeed, the widespread discharge of untreated or inadequately treated wastewater into streams, as well as the presence of rapidly evolving pollutants such as medicines, pesticides, and industrial chemicals, worsen the contamination of freshwater resources. It has been found that the quantities of ibuprofen and naproxen in fresh water and wastewater range from a hundred times of ng/L[1] to ten times of mg/L respectively. Consequently, the consumption of untreated or insufficiently treated water poses a number of significant dangers to human beings. Methods such as biodegradation, coagulation and flocculation, ozonation, and membrane filtration have all been tried in an effort to find a solution to this problem. However, none of these methods have been successful. Toxic by-product generation and high maintenance and servicing costs are just two of the many evident obstacles that limit the widespread implementation of these approaches. [4–6]. Figure 3.1 shows different types of advanced oxidation process for water treatment using TiO$_2$ doped cellulose or carbon matrix.

Photocatalytic degradation processes have been shown to be potent for the elimination of organic [7, 8] or inorganic toxins [9], along with microbes, such as bacteria, as well as fungi. The heterogeneous photo-catalysts are predicated on semiconductor materials, such as metal oxides [10, 11] and sulphides [11]. The photo-catalytic activity is summarised in Figure 3.2, in which it is shown that the efficiency to form

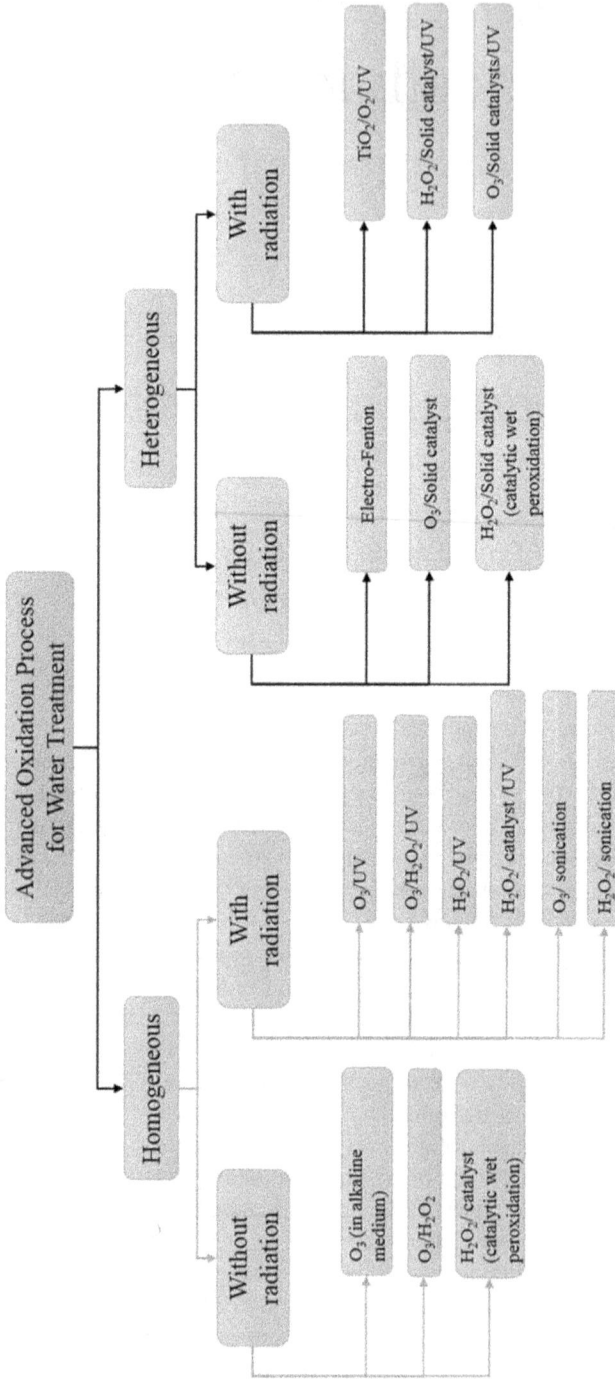

**FIGURE 3.1**  Classification of Advanced Oxidation Process for Water Treatment Using TiO$_2$ Doped Cellulose or Carbon Matrix.

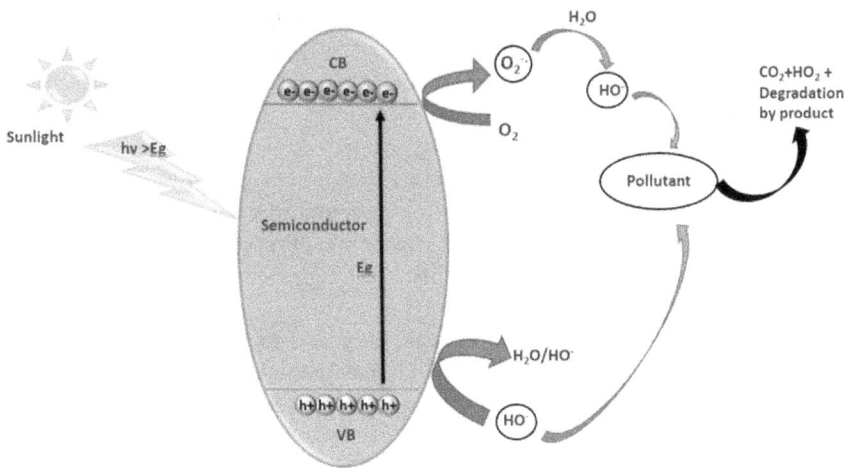

**FIGURE 3.2**  Photocatalytic Activity for Degradation of Organic Compounds.

OH- radicals is associated with the band gap of the photo-catalyst as well as the migration of electrons and holes from valence band to conduction band.

The efficiency of photo-induced electrons and holes is diminished when it is simple for them to recombine. Although numerous materials have been employed in photocatalysis, titanium dioxide (TiO$_2$) remains the most preferred and commonly utilised substance because of its minimal price, low cytotoxicity, great oxidising power, chemical and biological barrier properties, economic viability, poor solubility, and lengthy corrosion resilience. Nevertheless, the photocatalytic activity of TiO$_2$ is limited because of its high band-gap value (3.2 eV for anatase forms and 3.0 eV for rutile frameworks, correspondingly) as well as its poor quantum yield, which is caused by the substantial electron-hole recombination [12]. Due to this, TiO$_2$ can capture just a small portion (about 5%) of the solar spectrum of light, which results in good performances when employing UV light (wavelength 15,400 nm). Unfortunately, the activity is severely reduced when exposed to solar light.

It is quite clear that the utilisation of solar light as opposed to ultraviolet light for the photo-electro-catalytic water treatments presents an intriguing contest from a monetary as well as environmental perspective. It may be possible to improve the efficacy of water treatment by eliminating the need for electricity and chemicals, simplifying the facilities used (thus reducing the number of stages involved in the process), and utilising the abundant solar energy that is available globally. These considerations should enable the implementation of these know-hows in unindustrialised nations, allowing people to have access to fresh water.

Consequently, the study of innovative synthesis methods for the amalgamation of photo-catalysts based on TiO$_2$ to improve their opto-electrical attributes, resulting in exceedingly active photo-catalysts under the emissions from solar light, is an actively researched subject. For example, the incorporation of non-metal (N, S) or metal (Ag, Pd, etc.) doped agent into TiO$_2$ [13, 14] doping agents shrinks the band-gap by forming fresh hybridised conditions that gives considerable visible absorptivity of

Titania. The application of sensitisers encourages the immediate uptake of visible sunlight by it, which in turn triggers the redox cycle and releases electrons from $TiO_2$ into the surrounding environment. Even so, these techniques have a few disadvantages, such as the difficulties involved of reaching suitable concentration levels of dopants, the inadequate stability of the reactive groups while being exposed to irradiation and the emergence of defects that work as centres for the conjugation of photo-induced carriers ($e^2$ and $h^1$) [15]. It was found that combining the $TiO_2$ and ZnO produced hetero-junctions that had a lower rate of recombination and increased electron hole lifespan. This was accomplished through the transfer of electrons from ZnO towards $TiO_2$ [16].

In a similar vein, there is a considerable emphasis in the utilisation of biopolymers for the purpose of stabilising $TiO_2$ nanoparticles. The biopolymers are extracted from biomass or after conversion to carbon; heterogeneous catalysts can be developed from $TiO_2$ which can initiate photocatalytic degradation of organic pollutants. Cellulose is the abundantly available biopolymer that can be generated directly from biomass at lower cost. Earlier, $TiO_2$/microcrystalline cellulose based-composites were developed for the photo-degradation of colours from wastewater [17]. Based on the cellulosic pre-treatments, the emergence of novel surface active groups are developed. This influences the interaction between the phases, scattering, and biochemical composition of the ionic species and therefore, at the end, distinct photo-electro-catalytic performance. Nonetheless, the yield of increases in all circumstances when compared to pure phases, indicating the synergistic effect of phases, once more. These interactions also control the changes that take place throughout the carbonisation procedure; the crystalline phase and character of the catalyst phases are determined by the ratio of cellulosic substrate to $TiO_2$ and the temperatures at which the carbonisation takes place [17]. In this scenario, biochar-$TiO_2$ composites contribute to an increase in the overall yield of the pure phases. It is possible to develop composite substrate that can be employed as a photo-catalyst with modified physio-chemical properties as well as amended performance as well as stability through the utilisation of cellulosic materials, which are generally procured from agricultural wastes and the chemical modification of precursor chemicals or carbon/biochar derivatives.

It has been emphasised that adequate uptake of contaminants into the carbonaceous sorbent may limit the diffusion of the catalyst and hence may impair the overall process [18, 19]. A number of different efforts have claimed that it can be used (activated carbon (ACs)) as a substrate for the semiconductors. ACs supports the titanium oxide ($TiO_2$), which, when combined with the other mediums, has the potential to produce extraordinary outcomes [19]. These include its capability of rapidly absorbing contaminants as well as its great absorption ability as a result of its large surface active area and higher porosity [18]. Furthermore, due to a high capacity to capture liquid, ACs could decrease the permeation of ultra-violet lights into tiny places, and it may end up limiting the pollutants inside the pores, preventing them from being able to disperse into the exterior surface for additional reaction with the –OH radical [20, 21]. In addition, many types of pollutants, such as phenol, have the potential to undergo polymerisation on the carbon surface of the ACs, which results in permanent absorption.

## 3.2 SYNTHESIS OF TIO$_2$/CELLULOSE AND ITS DERIVATIVES BASED COMPOSITES FOR WASTEWATER PURIFICATION

Titania/cellulose composite materials can be obtained using a wide variety of methods, each of which can be subdivided into one of three types according to the role that cellulose plays in the composite. Therefore, cellulose can serve as a (a) support to distribute the coating of titania (b) a compensatory or persistent template to generate porous type titania, and (c) carbonaceous substrate to produce titania infused carbon composites. All of these roles are important for the production of titania/carbon composite materials. The composite properties vary a lot based on different sources of biomass matrix. Based on origin, biopolymers can be classified into different types. Figure 3.3 shows the sources of biomass and Figure 3.4 illustrates different types of biopolymers which have been extensively used for composite fabrication.

### 3.2.1 TIO$_2$ COATING OVER THE CELLULOSIC FILAMENTS

The production of titania onto the surface of cellulosic filaments is sometimes not effective enough to bind the TiO$_2$ which provides instable structure with non-durable bio-composites. It is possible to increase the durability of titania-based coating by introducing more cross-linking molecules onto the cellulosic matrix [22]. However, this will result in alterations to the composite architecture. Titania was able to be bound to the cellulose fibre surface through electrostatic interactions with –COOH groups.

In addition, the electrochemical properties of cellulose can be used to immobilise other preferred compounds of significance, like laccase, along with TiO$_2$ in order

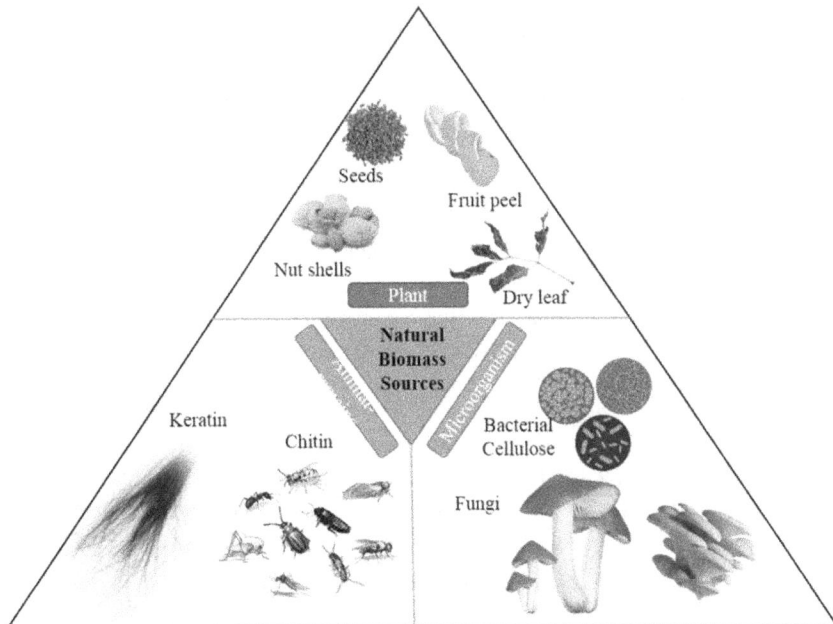

**FIGURE 3.3** Different Sources of Biomass Residues.

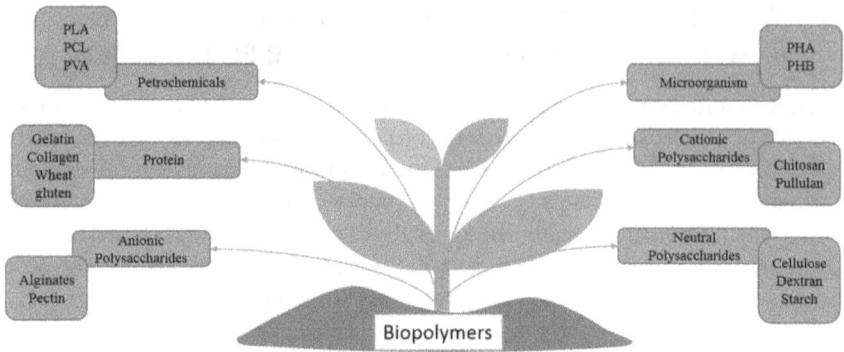

**FIGURE 3.4**  Classification of Biopolymers Based on Origin.

to acquire advanced materials which incorporate the bio-catalytic properties of laccase with the photo induced catalytic property of $TiO_2$ for enhanced degradation of dyes [23]. The nucleation and growth mechanism of $TiO_2$ particles over cellulosic filament are illustrated by Figure 3.5. Figure 3.6 illustrates the bonding mechanism of $TiO_2$ with the cellulosic matrix.

### 3.2.2  CELLULOSE DERIVED CARBON AS SUPPORT MATRIX OF $TiO_2$

The preparation of carbon based $TiO_2$ composite for photo induced catalytic degradation has also received a lot of interest. Carbonaceous components improve the sorption of the targeted contaminant over the surface of carbon, accompanied by a translocation across the interphase towards the $TiO_2$ phases, giving a synergistic interaction that uses photo induced electron/hole couples more effectively.

For the preparation of carbon/$TiO_2$ composites with strong photocatalytic efficiency, several types of carbon elements have been employed. They range from traditional carbons such as activated carbon (AC), graphite, carbon black, or other nano-dimensional carbons such as carbon nano fibres (CNFs), carbon nanotubes, 2D graphene, nano-horns, 3D fullerenes, or carbon aerogel [24]. However, it is important to take into consideration the usage of inexpensive precursors that are renewable, biodegradable, and beneficial to the environment. In this context, there is a growing interest in the fabrication of $TiO_2$ nanomaterials that are anchored on natural polymers like cellulose. However, in order to acquire the charcoal phase from the cellulosic precursor, an additional step of carbonisation is required, which is otherwise quite comparable to the procedures outlined for the preparation of cellulosic fibre-templated composites. Therefore, cellulose is capable of functioning as both an architectural framework and a source of carbon, with distinct hierarchical structure ranging from macroscopic to nano-scale levels. These composites demonstrated improved catalytic activity [25].

### 3.2.3  POROUS $TiO_2$ BASED COMPOSITES USING CELLULOSIC FILAMENT AS SACRIFICIAL TEMPLATES

Utilising physically insoluble cellulosic filaments as moulds for reproduction is one method for employing cellulose as a bio-template during the production of hybrid,

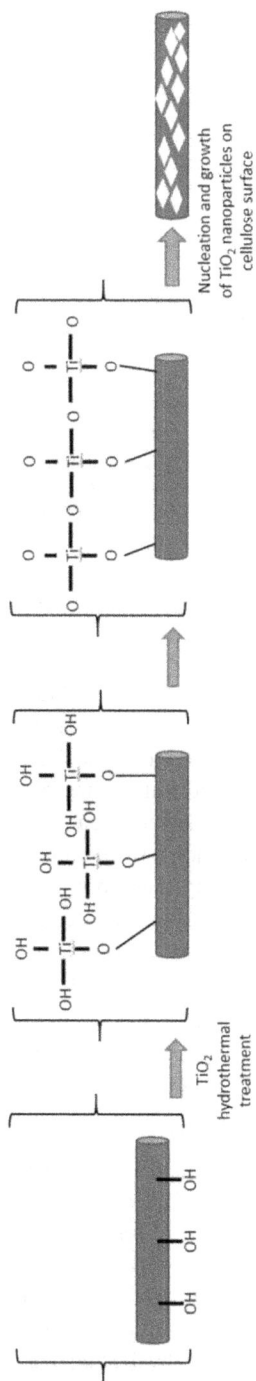

**FIGURE 3.5** Nucleation and Growth of TiO$_2$ Nanoparticles over Cellulosic Filament.

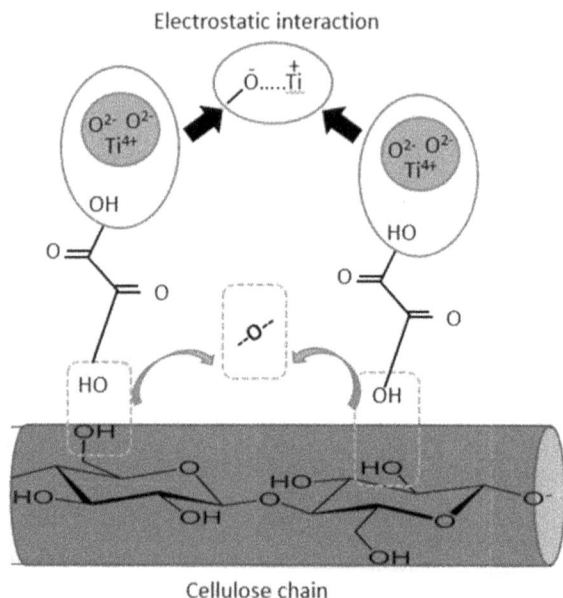

**FIGURE 3.6**   Nucleation and Growth of TiO$_2$ Nanoparticles over Cellulosic Filament.

porous nano-composite. This technology creates leverage for the variability and endurance of organic fibres, and as a result, it provides new chances for tuning the architectural, biochemical, and physical features of the products. The synthesis of titania with uncommon morphological features, such as fibres, needle, and cylinders that have previously been documented with other methodologies is now possible to use with cellulose as a template. Earlier research has used nanocrystalline cellulose as a component in the production of titanium dioxide [26]. In addition, the technique can be helpful in developing meso-porous TiO$_2$ nanomaterials which is necessary for photocatalytic treatment of wastewater [27]. The preparation of hierarchical nano-structured titania powder was accomplished through the use of one-pot hydrolysis of titanium dioxide precursors with CNCs [28].

## 3.3   SYNTHESIS OF TIO$_2$/CARBON AND ITS DERIVATIVES BASED COMPOSITES FOR WASTEWATER PURIFICATION

Formulation of TiO$_2$/activated carbon (ACs)-based catalyst can be accomplished by a variety of techniques, including chemical vapour deposition (CVD), precipitation, hydrothermal, pyrolysis, dip coating, sol-gel, and hydrolysis methods [29]. Nevertheless, the screening process that is employed for the choice of a compatible impregnation technique is solely dependent on the support materials and targeted pollutant [30]. Physicochemical characteristics of TiO$_2$/ACs substrate have a large influence on the architecture of the solid support and rely on production methods, such as heat treatments. The primary benefits of adopting physical techniques are that they are straightforward and inexpensive, and their application in commercial settings is already established for photo-catalysts with the desired capabilities. TiO$_2$/ACs combination was also conducted by employing the standard wet procedures, but the discrepancies at the levels of lattice reduced the requisite effectiveness of isolation and transportation of photo-generated charge carriers (electrons and electrons) [31, 32].

It would appear that the adhesion of ACs interface to TiO$_2$ particles is significant for the enhancement of photo and electro-catalytic activity, as well as for the practical uses of hybrid systems. The wet procedure of synthesis is indicated since it is necessary to improve the anchoring of TiO$_2$ onto ACs [32]. Because the physically stable TiO$_2$/ACs hybrid is in conflict with the hydrostatic shearing approach, surface interactions and porous texture of AC can have adequate repercussions on distributing the TiO$_2$ over the AC matrix.

## 3.4 PHOTOCATALYTIC DEGRADATION PERFORMANCE BY TIO$_2$ DOPED CELLULOSIC MATRIX FOR WATER TREATMENT

The utilisation of cellulose/TiO$_2$ composite for photocatalytic wastewater treatment has a number of benefits, including but not limited to the following:

1. The presence of hydrophilicity inside the cellulosic chain can encourage the nucleation as well as the development of inorganic phase like titania, which ultimately results in good porosity and a reduction in the fouling of the generated nano composite [33].
2. Improve the photocatalytic properties by controlling the speed of hydrolysis, development of the metal oxide/cellulose, and poly-condensation [34].
3. To ensure that the photo-catalyst is distributed evenly throughout the cellulosic matrix as a consequence of the interaction that occurs at the interface between the two phases [35].
4. The presence of cellulose, which provides both a mechanical support and an appealing matrix component for the dispersion of photo-catalysts [36].
5. Enhancing the optical characteristics of the cellulose-based membrane matrix by increasing the intensity of the irradiation from the source of light inside the composite of the membrane, which will then lead to an enhancement in the distribution of electrons and the transfer of electrons to the photo-catalyst surface [36].
6. Changing the optoelectronic properties of TiO$_2$ so that it is less sensitive to UV light and more sensitive to visible light by modifying the surface of TiO$_2$, by combining it with cellulosic substrate [37]. These adjustments are helpful in lowering the rate at which electrons and holes recombine, which indicates an enhanced quantum efficiency of the photocatalytic process. Constraining recombination by raising charge transfer and consequently the effectiveness of the photocatalytic process is one of the advantages of surface treatment. Another benefit is raising the wavelength response spectrum so that the catalyst can be energised in the visible region of light, and a third benefit is shifting its selectivity or yields of a specific material.
7. Improving the thermo-stability of TiO$_2$ so that it can be used for treatments at higher temperatures while also preventing the membrane from being destroyed by ultraviolet light [38].

Titania-cellulose hybrid composite has been shown to be effective in a variety of photocatalytic application areas, including the elimination of stains and the emergence of a self-cleaning exterior, the destruction of organic contaminants, the neutralisation of odours caused by bacteria, protection from ultraviolet light, resistance to flame, and removal of nitrogen oxides.

A significant amount of research has been put into investigating whether or not it is possible to immobilise or deposit cellulose based $TiO_2$ nano composites inside the membranes for wastewater purification. For example, cellulose-based $TiO_2$ composites were developed by immobilising titania in a cellulosic matrix [39]. These composites showed high catalytic performance in the breakdown of phenol when they were exposed to UV light. Researchers developed a highly recyclable catalyst by synthesising a flexible highly porous $TiO_2$ micro-sphere/cellulose acetate (CA) hybrid film [40].

## 3.5 PHOTOCATALYTIC DEGRADATION PERFORMANCE BY TIO$_2$ DOPED CARBONACEOUS SUBSTRATE FOR WATER TREATMENT

It is vital to analyse the sorption efficiency for $TiO_2$/ACs, despite the fact that the photo-catalytic degradation performance is the most intriguing and important of the two. Regrettably, the research on $TiO_2$/ACs does not frequently publish findings from investigations of this kind. It was only mentioned in a handful of studies that dealt with $TiO_2$/ACs or mixes of $TiO_2$/ACs [41, 42]. Figure 3.7 displays the layout of $TiO_2$-based ACs supported by photo-reactors which can be used for water treatment.

As a result of the pores being blocked, the sorption efficiencies of $TiO_2$/ACs can be slightly lower than that of pure ACs. This is something that is quite understandable. The extent to which the sorption capacity has been reduced can perhaps be

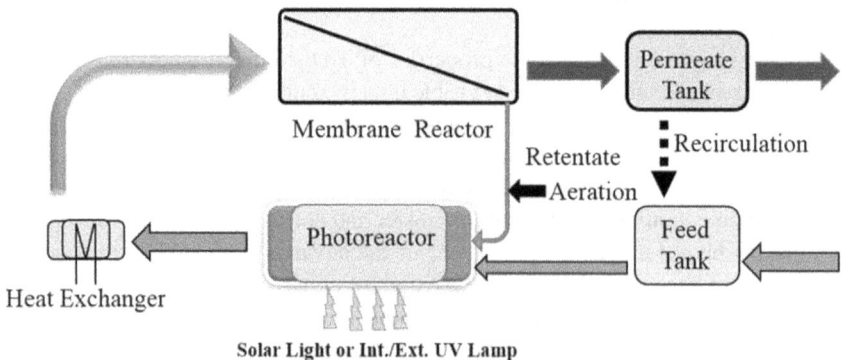

**FIGURE 3.7** Basic Layout of Photocatalytic Reactor Containing $TiO_2$-Based ACs as Membrane for Water Treatment.

determined by quantifying the changes that have occurred in the BET surface area following the TiO$_2$ deposition.

When conducting studies to assess the efficacy of a TiO$_2$/ACs composite for the removal of contaminants, there are a few ways that may be utilised to differentiate between the impacts of adsorption and catalytic degradation. This is necessary due to the fact that ACs is an excellent adsorbent. After reaching an equilibrium of pollutant sorption in the dark, the approach that is most frequently used is called the batch photo-catalytice experiment. During this experiment, one looks for a consistent reduction in the amount of contaminants present while they are exposed to UV light. The degree of the catalytic impact can also be determined by conducting a batch test that involves repeated spiking of the targeted pollutant in addition to measurements of its elimination over the course of multiple cycles and mass balancing. Another method is to conduct a continuous flow experiment with TiO$_2$/ACs that is enclosed inside a photocatalytic reactor (for example, with an appropriate membrane) and then measure the effectiveness with which the contaminant is removed while the system is in a stable state. The identification of intermediates and by-products could provide definitive proof that contaminants have been removed using a photo-catalytic degradation pollutant from wastewater. The photocatalytic mechanism of ACs-supported TiO$_2$ composite is shown by Figure 3.8.

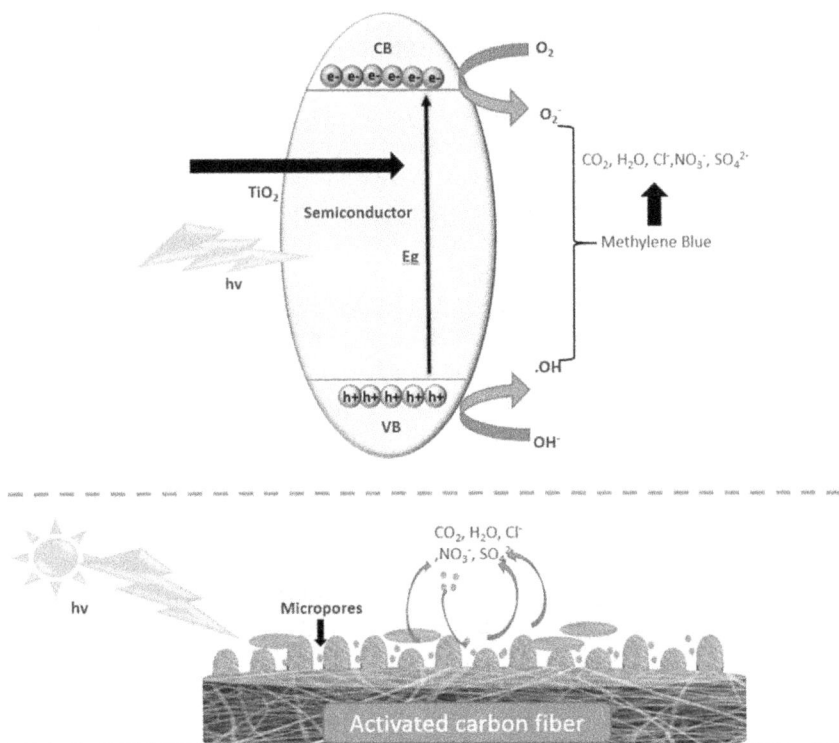

**FIGURE 3.8** Mechanism of Photo-Degradation of Organic Pollutant Using TiO$_2$/Acs.

## 3.6 RECENT ROADMAP, CHALLENGES AND FUTURE PERSPECTIVE OF TIO₂ DOPED BIOPOLYMER COMPOSITES

This approach emphasises current breakthroughs in the application of ligno-cellulosic biomass-based biopolymer or its carbon derivative-based photo-catalysts for the degradation of contaminants from waste effluents. The work is based on the substantial number of relevant articles that have been published up to this point. Photo-catalysts that are derived from biomass-based carbon offer a number of beneficial properties, including the fact that they are non-toxic, inexpensive, plentiful, and very permeable. The frameworks of these materials that make up activated carbon can lead to exceptional sorption capacities as well as enhanced oxidation ability for a wide variety of organic contaminants. Nevertheless, there are a number of loopholes in the research that require additional attention. Carbonisation at elevated heat is an essential stage, and as a result, the processing of biomass-based carbon requires a significant amount of energy. As a direct consequence of this, the rising price of generating carbon continues to be a barrier that prevents scale-up. Second, there have been no studies conducted to this day that concentrate on the financial implications of utilising biomass-derived carbon-modified semiconductors in any capacity. Furthermore the dispersion of $TiO_2$ in cellulosic matrix is still unclear. The economic viability of the project as well as its effects on the surrounding environment are two important aspects that need careful consideration. In addition, the majority of recent investigations are only conducted on a small scale in the laboratory. The biomass can be used as a source of reduced-cost carbon constituents for applications related to environmental protection. As a result, the manufacture of carbon from biomass at a big scale needs to be further improved. Another thing to consider is that the properties of the carbon materials that are created can be significantly influenced by the biomass resources that are used, which can have a wide variety of properties, as well as the circumstances in which they are synthesised. As a result, additional research is required to discover the optimum content of the original biomass as well as the appropriate synthesis parameters that match to that composition.

In subsequent studies, investigation will be required to explore the specific links between the topologies and physiochemical properties of carbon material and the photochemical performances of these materials.

Without a doubt, the conversion of naturally plentiful biomass into carbon-enriched materials can also help to the overall effort to reduce waste on a global scale. As a result, it is reasonable to anticipate that significant progress will be made in the field of environmental restoration in the years to come. It is expected that the readership of this work will be informed about acceptable, simple, inexpensive, and ecologically friendly methods comprising the usage of biomass matrix containing cellulose as the basic building block. This will enable the ongoing expansion of excellent photo-catalysts for environmental remediation.

## 3.7 CONCLUSION

Cellulose-titanium dioxide composites are exciting new materials that have the potential to be used in heterogeneous photo-catalysis for the purification of water.

Various techniques for isolating nanocellulose result in the production of distinct types of nanocellulose-based materials, such as nanocrystalline cellulose (NCC), microcrystalline cellulose (MFC), and bacterial cellulose (BNC), each of which has its own unique structure, set of features, and set of uses. Because of its highly consistent and crystallinity index, cellulose must be pre-treated or functionalised before it can be used for the production of cellulosic substrate based titania composites. This is one of the most important steps in the process. Therefore, the methods that can be used on cellulose include those that involve its solubilisation in water-based or protic fluids, its dissolution in water immiscible environments, and its dispersion by physical or chemical interventions. These techniques can stimulate not only modifications in the configuration of the nano or micro dimensional cellulosic materials (nanocrystalline-CNCs or microcrystalline MFC) but also the inclusion of biochemical features and functionality, which favour the strong interaction with TiO$_2$ during the fabrication of the composite materials.

When it comes to the production of cellulose-titanium composite, cellulose can serve as a support for a layer or coating of titanium dioxide (TiO$_2$) that might be coated or formed in situ over it using a variety of processes. Utilising cellulose as just a protective or persistent template is another method that can be used to produce porous titania. When compared to the porous features achieved using the first technique, improved porosity was observed. Last but not least, cellulose can be carbonised once it has been coupled or combined with TiO$_2$ to generate carbon/TiO$_2$ composite, which can then be used as a carbon source.

The incorporation of bi-functional cellulose or its carbon matrix, together with titania, has exhibited improved photocatalytic performance. This is due to the integration of numerous factors, including the following: (1) inhibition of electro-hole recombination process, (2) enhanced sorption behaviour, (3) improved photo-stability, (4) amplified light absorption, as well as (5) synergistic effects of pollutant eradication.

The photo-induced catalytic oxidation of organic pollutants (AOP) and carbon adsorption are two critical mechanisms in advanced wastewater purification methods and restoration; the TiO$_2$/activated carbon (ACs) hybrid offers both of these processes. Another distinguishing property of the composites is that they might very well be designed to successfully address particular groups of pollutants by varying their bi-functional behaviour of adsorption/photocatalytic degradation, such as by altering the loading of TiO$_2$ or the type of ACs-based support based on lignocellulosic materials types. It is advised that the sorption and photocatalytic degradation processesbe coupled with an efficient particle separation technology such as a filter membrane unit. It is proposed that a hybrid system could exist, consisting of a reaction chamber using a membrane and a photo-reactor containing diffused TiO$_2$/ACs. Prospective interactions with biological processes to improve contaminant mineralisation or with additional water purification methods to reduce total energy and chemical usage are indeed considered. This emerging class of visible-light sensitive TiO$_2$/ACs hybrids may offer an exciting opportunity for its ongoing deployment in the removal of contaminants with concurrent in-line replenishment utilising either UV light or solar light. This would be a win-win situation.

## ACKNOWLEDGMENTS

The authors are thankful for the funding provided by Malaysian Joint Research Scheme-ST 077–2022, Interdisciplinary Research IIRG003A-2022IISS and International Grant ICF 023–2022 and ICF 080–2021 under University of Malaya, Kuala Lumpur 50603, Malaysia.

## REFERENCES

[1] Hurt, R. H., Monthioux, M. and Kane, A. 2006. Toxicology of carbon nanomaterials: status, trends, and perspectives on the special issue. *Carbon*, 44(6), 1028–1033.

[2] Subramanian, V., Zhu, H., Vajtai, R., Ajayan, P. M. and Wei, B. 2005. Hydrothermal synthesis and pseudocapacitance properties of $MnO_2$ nanostructures. *Journal of Physical Chemistry B*, 109(43), 20207–20214.

[3] Seiler, W. and Crutzen, P. J. 1980. Estimates of gross and net fluxes of carbon between the biosphere and the atmosphere from biomass burning. *Climatic Change*, 2(3), 207–247.

[4] Philip, M. F., Niwton, L. and Fernondo, M. F. 1980. Rainforest burning and the global carbon budget: Biomass, combustion efficiency, and charcoal formation in the Brazilian Amazon. *Journal of Geophysical Research: Atmospheres*, 98, 16733–16743.

[5] Kuzyakos, Y., Subbotino, I., Chen, H., Bagomolova, I. and Xu, X. 2009. Black carbon decomposition and incorporation into soil microbial biomass estimated by 14C labelling. *Soil Biology and Biochemistry*, 41(2), 210–219.

[6] Demirbas, A. 2001. Carbonization ranking of selected biomass for charcoal, liquid and gaseous products. *Energy Conversion and Management*, 42(10), 1229–1238.

[7] Bailón-García, E., Elmouwahidi, A., Carrasco-Marín, F., Pérez-Cadenas, A. F. and Maldonado-Hódar, F. J. 2017. Development of Carbon-$ZrO_2$ composites with high performance as visible-light photocatalysts. *Applied Catalysis B: Environment*, 217, 540–550. https://doi.org/10.1016/j.apcatb.2017.05.090.

[8] Shaban, Y. A., El Maradny, A. A. and Farawati, R. K. A. 2016. Photocatalytic reduction of nitrate in seawater using C/$TiO_2$ nanoparticles. *Journal of Photochemistry Photobiology A Chemistry*, 328, 114–121. https://doi.org/10.1016/j.jphotochem.2016.05.018.

[9] Singh, R., Verma, K., Patyal, A., Sharma, I., Barman, P. B. and Sharma, D. 2019. Nanosheet and nanosphere morphology dominated photocatalytic & antibacterial properties of ZnO nanostructures. *Solid State Science*, 89, 1–14. https://doi.org/10.1016/j.solidstatesciences.2018.12.011.

[10] Chang, X., Li, Z., Zhai, X., Sun, S., Gu, D., Dong, L., et al. 2016. Efficient synthesis of sunlight-driven ZnO-based heterogeneous photocatalysts. *Materials Design*, 98, 324–332. https://doi.org/10.1016/j.matdes.2016.03.027.

[11] Hosseinpour, Z., Arefinia, Z. and Hosseinpour, S. 2018. Cu doped cubic pyrite-type $CoS_2$ ball like superstructures as heterogeneous photocatalyst. *Materials Chemistry Physics*, 220, 426–432. https://doi.org/10.1016/j.matchemphys.2018.09.004.

[12] Parnicka, P., Mazierski, P., Grzyb, T., Lisowski, W., Kowalska, E., Ohtani, B., et al. 2018. Influence of the preparation method on the photocatalytic activity of Nd-modified $TiO_2$. *Beilstein Journal of Nanotechnology*, 9, 447–459. https://doi.org/10.3762/bjnano.9.43.

[13] Mao, H., Fei, Z., Bian, C., Yu, L., Chen, S. and Qian, Y. 2019. Facile synthesis of high-performance photocatalysts based on Ag/$TiO_2$ composites. *Ceramic International*, 45, 12586–12589. https://doi.org/10.1016/j.ceramint.2019.03.109.

[14] Zarepour, M. A. and Tasviri, M. 2019. Facile fabrication of Ag decorated $TiO_2$ nanorices: Highly efficient visible-light-responsive photocatalyst in degradation of contaminants.

*Journal of Photochemistry Photobiology, A Chemistry*, 371, 166–172. https://doi. org/10.1016/j.jphotochem.2018.11.007.

[15]  Zheng, X., Li, D., Li, X., Chen, J., Cao, C., Fang, J., et al. 2015. Construction of ZnO/TiO₂ photonic crystal heterostructures for enhanced photocatalytic proper- ties. *Applied Catalysis B: Environment*, 168–169, 408–415. https://doi.org/10.1016/j. apcatb.2015.01.001.

[16]  Kitano, M., Funatsu, K., Matsuoka, M., Ueshima, M. and Anpo, M. 2006. Preparation of nitrogensubstituted TiO₂ thin film photocatalysts by the radio frequency magnetron sput- tering deposition method and their photocatalytic reactivity under visible light irradiation. *Journal of Physical Chemistry B*, 110, 25266–25272. https://doi.org/10.1021/jp064893e.

[17]  Hamad, H., Castelo-Quibe´n, J., Morales-Torres, S., Carrasco-Marı´n, F., Pe´rez- Cadenas, A. F. and Maldonado-Ho´dar, F. J. 2018. On the interactions and synergism between phases of carbon-phosphorus-titanium composites synthetized from cel- lulose for the removal of the Orange-G dye. *Materials* (Basel), 11, 1766. https://doi. org/10.3390/ma11091766.

[18]  Kumar, S., Kumar, B., Surender, T. and Shanker, V. 2014. g-C₃N₄/NaTaO₃ organic– inorganic hybrid nanocomposite: High performance and recyclable visible light driven photocatalyst. *Materials Research Bulletin*, 49, 310–318.

[19]  Gamage McEvoy, J., Cui, W. and Zhang, Z. 2014. Synthesis and characterization of Ag/ AgCl-activated carbon composites for enhanced visible light photocatalysis. *Applied Catalysis B: Environmental*, 144, 702–712.

[20]  Yun, J., Kim, H. I. and Lee, Y. S. 2013. A hybrid gas-sensing material based on porous carbon fibers and a TiO₂ photocatalyst. *Journal of Materials Science*, 48, 8320–8328.

[21]  Adeli, B. and Taghipour, F. 2013. A review of synthesis techniques for gallium-zinc oxynitride solar-activated photocatalyst for water splitting. *ECS Journal of Solid State Science and Technology*, 2(7), Q118–Q126.

[22]  Karimi, L., Mirjalili, M., Yazdanshenas, M. E. and Nazari, A. 2010. Effect of nano TiO₂ on selfcleaning property of cross-linking cotton fabric with succinic acid under UV irradiation. *Photochemistry Photobiology*, 86, 1030–1037. https://doi. org/10.1111/j.1751–1097.2010.00756.x.

[23]  Li, G., Nandgaonkar, A. G., Wang, Q., Zhang, J., Krause, W. E. Wei, Q., et al. 2017. Laccase-immobilized bacterial cellulose/TiO₂ functionalized composite membranes: evaluation for photo- and bio-catalytic dye degradation. *Journal of Membrane Science*, 525, 89–98. https://doi.org/10.1016/j.memsci.2016.10.033.

[24]  Leary, R. and Westwood, A. 2011. Carbonaceous nanomaterials for the enhancement of TiO₂ photocatalysis. *Carbon N. Y.*, 49, 741–772. https://doi.org/10.1016/j.carbon. 2010.10.010.

[25]  Qi, D., Li, S., Chen, Y. and Huang, J. 2017. A hierarchical carbon@TiO₂@MoS2 nano- fibrous composite derived from cellulose substance as an anodic material for lithium- ion batteries. *Journal of Alloys Compounds*, 728, 506–517. https://doi.org/10.1016/j. jallcom.2017.09.018.

[26]  Zhou, Y., Ding, E. Y. and Li, W. D. 2007. Synthesis of TiO₂ nanocubes induced by cel- lulose nanocrystal (CNC) at low temperature. *Materials Letters*, 61, 5050–5052. https:// doi.org/10.1016/j.matlet.2007.04.001.

[27]  Xue, J., Song, F., Yin, X. W., Zhang, Z. L., Liu, Y., Wang, X. L., et al. 2017. Cellulose nanocrystal-templated synthesis of mesoporous TiO₂ with dominantly exposed (001) facets for efficient catalysis. *ACS Sustainable Chemical Engineering*, 5, 3721–3725. https://doi.org/10.1021/acssuschemeng.7b00341.

[28]  Chen, T., Wang, Y., Wang, Y. and Xu, Y. 2015. Biotemplated synthesis of hierarchi- cally nanostructured TiO₂ using cellulose and its applications in photocatalysis. *RSC Advances*, 5, 1673–1679, https://doi.org/10.1039/c4ra13955k.

[29]   Wang, Y., Ren, P., Feng, C., Zheng, X., Wang, Z. and Li, D. 2014. Photocatalytic behavior and photo-corrosion of visible-lightactive silver carbonate/titanium dioxide. *Materials Letters*, 115, 85–88.

[30]   Li, S. and Ye, G. 2012. Photocatalytic degradation of formaldehyde by $TiO_2$ nanoparticles immobilized in activated carbon fibers. *Advanced Materials Research*, 482–484, 2539–2542.

[31]   Sun, S. M. 2012. Enhancement of $TiO_2$ behaviour on photocatalytic oxidation of MO dye using $TiO_2$/AC under visible irradiation and sunlight radiation. *Applied Mechanics and Materials*, 485, 161.

[32]   Riaz, N., Chong, F. K., Dutta, B. K., Man, Z. B., Khan, M. S. and Nurlaela, E. 2012. Photodegradation of orange II under visible light using Cu-Ni/$TiO_2$: Effect of calcination temperature. *Chemical Engineering Journal*, 185–186, pp. 108–119.

[33]   Gutierrez, J., Fernandes, S. C. M., Mondragon, I. and Tercjak, A. 2013. Multifunctional hybrid nanopapers based on bacterial cellulose and sol-gel synthesized titanium/vanadium oxide nanoparticles. *Cellulose*, 20, 1301–1311. https://doi.org/10.1007/s10570-013-9898-2.

[34]   Su, X., Liao, Q., Liu, L., Meng, R., Qian, Z., Gao, H., et al. 2017. $Cu_2O$ nanoparticle-functionalized cellulose-based aerogel as high-performance visible-light photocatalyst. *Cellulose*, 24, 1017–1029. https://doi.org/10.1007/s10570-016-1154-0.

[35]   Moon, R. J., Martini, A., Nairn, J., Simonsen, J. and Youngblood, J. 2011. Cellulose nanomaterials review: Structure, properties and nano-composites. *Chemical Society Reviews*, 40, 3941–3994. https://doi.org/10.1039/c0cs00108b.

[36]   Snyder, A., Bo, Z., Moon, R., Rochet, J. C. and Stanciu, L. 2013. Reusable photocatalytic titanium dioxide-cellulose nanofiber films. *Journal of Colloid Interface Science*, 399, 92–98. https://doi.org/10.1016/j.jcis.2013.02.035.

[37]   Boufi, S., Abid, M., Bouattour, S., Ferraria, A. M., Conceição, D. S., Ferreira, L. F. V., et al. 2019. Cotton functionalized with nanostructured $TiO_2$-Ag-AgBr layer for solar photocatalytic Degradation of dyes and toxic organophosphates. *International Journal of Biological Macromolecules*, 128, 902–910. https://doi.org/10.1016/j.ijbiomac.2019.01.218.

[38]   Li, J. F., Xu, Z. L., Yang, H., Yu, L. Y. and Liu, M. 2009. Effect of $TiO_2$ nanoparticles on the surface morphology and performance of microporous PES membrane. *Applied Surface Science*, 255, 4725–4732. https://doi.org/10.1016/j.apsusc.2008.07.139.

[39]   Zeng, J., Liu, S., Cai, J. and Zhang, L. 2010. $TiO_2$ immobilized in cellulose matrix for photocatalytic degradation of phenol under weak UV light irradiation. *Journal of Physical Chemistry*, 114, 7806–7811.

[40]   Jin, X., Xu, J., Wang, X., Xie, Z., Liu, Z., Liang, B., et al. 2014. Flexible $TiO_2$/cellulose acetate hybrid film as a recyclable photocatalyst. *RSC Advances*, 4, 12640–12648. https://doi.org/10.1039/c3ra47710j.

[41]   Cordero, T., Duchamp, C., Chovelon, J.-M., Ferronato, C. and Matos, J. 2007. Influence of L-type activated carbons on photocatalytic activity of $TiO_2$ in 4-chlorophenol photodegradation. *Journal of Photochemistry and Photobiology A: Chemistry*, 191, 122–131.

[42]   Matos, J., Laine, J., Herrmann, J. M., Uzcategui, D. and Brito, J. L. 2007. Influence of activated carbon upon Titania on aqueous photocatalytic consecutive runs of phenol photodegradation. *Applied Catalysis B: Environmental*, 70, 461–469.

# 4 Electrospun Nanofibre/ Composite Membrane for Water Treatment

*Bagavathi Krishnan, Zaira Zaman Chowdhury,*
*Masud Rana, Ahmed Elsayid Ali, Rahman*
*Faizur Rafique, Amutha Chinnapan and*
*Md. Mahfujur Rahaman*

## 4.1 INTRODUCTION

Owing to inadequate management of water resources, mishandling, environmental destruction, pollution, and the massive expansion of the worldwide population, the global community is continuously enduring a severe problem with fresh water shortages [1, 2]. This is a problem that has been going on for quite some time. In light of the growing number of unfavourable consequences, several strategies for the management and purification of water resources are currently undergoing research and development [3, 4]. In comparison to other traditional approaches to water purification, the pressure-driven membranes technique is becoming increasingly popular all over the world and on a broader scale [5]. As a consequence of this, the membrane filtering approach has seen widespread use in the treatment of wastewater originating from a variety of sources [6–8]. During the process of separation, a membrane acts as a barrier and a selective boundary for a number of different biological and chemical entities. Membranes can be classified as structural or physical interphases. The evolution of membranes over time has led to their current categorisation as homogeneous or heterogeneous membranes; asymmetrical or symmetrical membranes, [9] and negative, positive, or neutrally charged membranes [10]. It is common ynthesi to categorise pressure-driving membranes in accordance with the sizes of their pores. For example, nanofiltration membranes (NF), ultrafiltration membranes (UF), microfiltration membranes (MF), and reverse osmosis membranes (RO) are all examples of such classifications [11]. It is normal ynthesi to employ these membranes (MF, UF, NF, and RO) in the process of successfully removing from water bodies suspended particles, proteins, macromolecules, and both multivalent and monovalent ions. The materials that make up a membrane are a significant factor in determining its properties and characteristics. Organic membranes and inorganic ceramic membranes that comprise bio or synthetic polymer or their composites are two types of membranes [12–15]. Because organic polymers possess excellent physical, mechanical, and chemical properties, they have found widespread use as components in a variety of commercial goods. Recent years have seen the development of several types of

DOI: 10.1201/9781003342830-4

membranes [16, 17]. It is estimated that cellulose acetate, polyethersulfone (PES), and polyvinylidene fluoride (PVDF), as well as their composites, hold the majority of the market share for polymeric membranes used in water filtration. Ceramic nano-particles, such as $Al_2O_3$, $SiO_2$, or $TiO_2$, are frequently used in the construction of inor-ganic membranes [17]. They are unlikely to find broad applicability due to their high price and fragility, which puts the economic viability of the membrane into question when it comes to the filtration and water purification process on an industrial level. This is in comparison to bio- and synthetic polymer-based membranes, which find widespread application. As a direct consequence of this, polymeric membranes are receiving a growing amount of attention all around the world.

Bio-polymer based membranes derived from cellulose are represented as one of the greatest potentially effective approaches towards the purification of water. This is due to the fact that they have a large filtration power for both contaminants and microbes. In addition, cellulose membranes have the capability of removing unwanted com-pounds from a variety of environments, including water, microbial fluids, biological contaminants, foodstuff, and air.

It has recently been demonstrated that bio-polymeric membranes like cellulose acetate (CA), cellulose nano-crystals (CNCs), and cellulose nano fibres (CNFs) exhibit an impressive capacity for filtration. Typically, bio-polymeric membranes that are dependent upon cellulosic fibres along with its variants remain manufactured from sustainable feedstock like industrial as well as agricultural wastes. Addition-ally, these types of membranes have been premeditated in order to solve problems that are associated with conventional polymeric membranes. This is as a result of the distinctive qualities of cellulosic membranes, which include a high capacity for adsorption; great resistance to chemicals, heat, and corrosion; superior production rates; and discharge of a minimal amount of harmful compounds [18, 19]. In point of fact, numerous bio-polymers are being developed for use in the actual filtration process that occurs in the environment.

The synthetic hydrophobic polymers, like polysulfone, polyacrylonitrile, polyeth-ersulfone, polysulfone, and poly(vinylidene fluoride) all have good physio-chemical and mechanical characteristics. However, in order for these polymers to be used in the production of a membrane for water filtration, they first ought to be functional-ised or otherwise modified [20]. As a consequence of this, hydrophilic polymers like cellulose acetate (CA) are ynthesi in the construction of all commercially available water filtration membranes, most notably those that are utilised in reverse osmosis (RO). Because the electrospun CA nanofibrous membrane does not possess anti-bacterial activity, the scope of its use is restricted to the purification of water. As a consequence of this, antibacterial agents need to be inserted inside the electrospun nanofibre membrane throughout the filtering process in order to eliminate microbes/germs from wastewater and to prevent the membrane from fouling [21, 22]. During the electrospinning process, the nanofibrous membrane may become contaminated with dirt and debris [23]. As a consequence of this, a wide variety of techniques for surface modification, such as coating, grafting, mixing, and interfacial polymerisa-tion, are ynthesi. Gentamicin, for instance, is ynthesi in the process of removing the microorganism from the water and reducing the amount of fouling that occurs on the membrane. During the process of water filtration, additives and antimicrobial compounds are used to modify the interface of the electrospun membrane in order to reduce the amount of biofouling [24]. The following is a list of commonly used materials ynthesi for the production of nanofibres via electrospinning, as shown in Figure 4.1.

However, fouling is a key concern for the advancement of bio/synthetic polymer derived membranes because of their intrinsic hydrophobicity. This favours the attachment of hydrophobic NOMs (natural organic matter), which can cause the membranes to get clogged [25, 26]. As a consequence of this, scientists are investigating a variety of methods to alter the properties of the polymeric membrane so that it becomes more hydrophilic [27]. In particular, it has been established that the use of a mixture of natural or synthetic polymers and nanoparticles is an efficient strategy for improving the hydrophilicity of the membrane used for water filtering [28]. In recent years, this approach has garnered a growing amount of interest for the development of antifouling membrane [29]. Previous research publications have investigated and reviewed $TiO_2$ nanocomposite-based polymeric membranes in great detail [30]. Zinc oxide (ZnO) nanoparticles, which cost one-fourth as much as titanium dioxide ($TiO_2$), have been proven in a number of studies to have equal physical and chemical properties to those of $TiO_2$. As a consequence of this, zinc oxide nanoparticles are regarded as a practical alternative to titanium dioxide nanoparticles in the manufacturing of fouling-resistant organic–inorganic composite membranes that are based on either a bio or synthetic polymeric matrix [31]. There have been a number of studies that have been released on the development of fouling resistant polymer–ZnO composite membranes for water treatment [31]. However, a comprehensive review on the performance appraisal of the electrospun bio and synthetic polymeric metal oxide (MO)-based composites is also still lacking in the literature, and additional research is required for advancement in this field.

The invention of an electrospun nanofibre-based membrane incorporating synthetic/biopolymer with photo-catalyst doping has shown to be a promising technique for the degradation of a pollutant that is present in wastewater by means of an improved oxidation process. Because of the increased surface area and porosity, there are more photocatalytic degradation sites that are easily accessible. This contributes to a higher level of electron-hole pair separation efficiency. It is possible to improve the effectiveness of synthetic polymer-based membranes used for water filtration by functionalising them with some type of biopolymer and making them commercially available. The creation of fouling-resistant electrospun nanofibrous membranes with photocatalytic capabilities that can be employed for the filtration of wastewater was the primary emphasis of this chapter. The accompanying section also discusses the overall performance, challenges, and future viewpoint of using selected metal oxide nanoparticles doped electrospun nanofibre-based composite membrane for the treatment of wastewater.

## 4.2 BASIC PRINCIPLE FOR ELECTROSPUN MEMBRANE FABRICATION FOR WATER TREATMENT

When electrospun, virtually every soluble chemical can always generate ynthesize with size varying from a few micrometres to tens of nanometers in length. This is

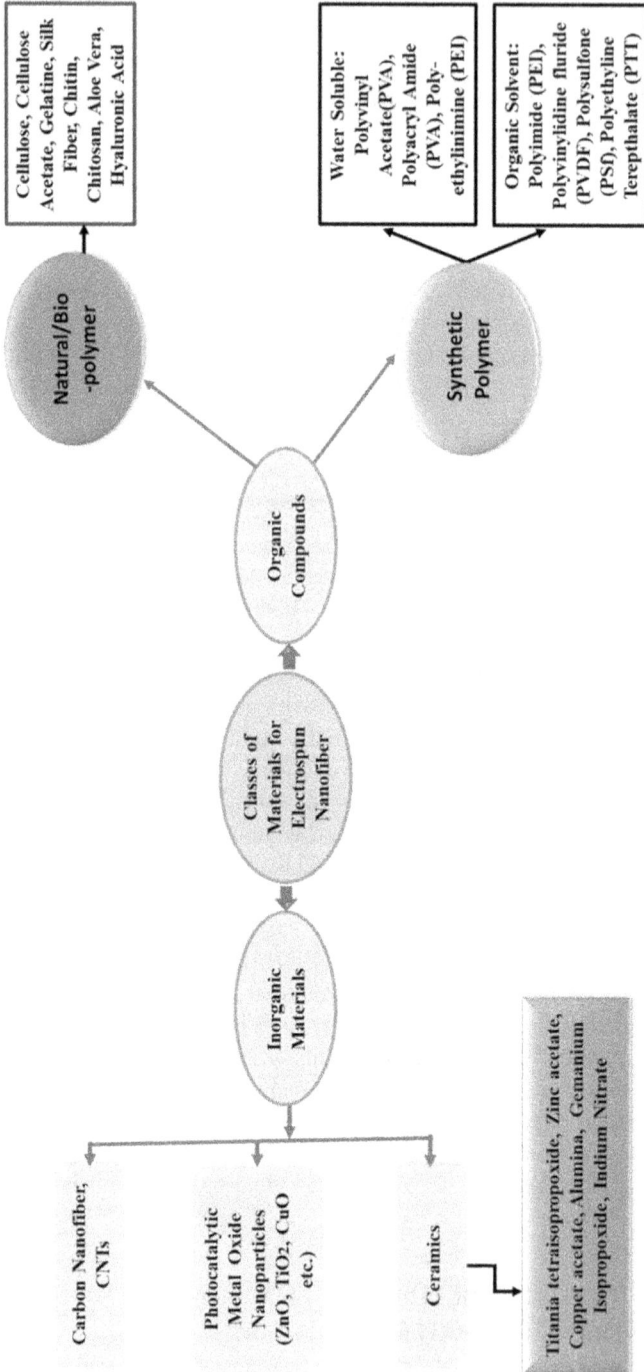

**FIGURE 4.1** Materials Used for Electrospun Membrane Fabrication.

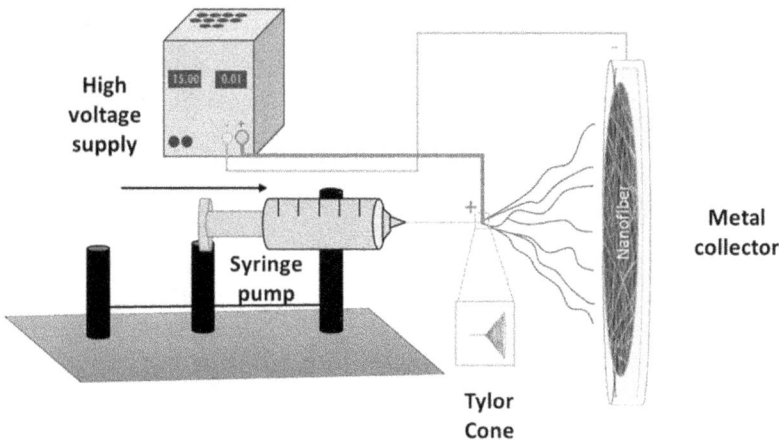

**FIGURE 4.2** Basic Layout of Electrospun Membrane Fabrication Process for Water Treatment

always the case. According to the researchers, approximately two hundred different bio or polymer blends of synthetic origin have been electrospun into long nanostructure of fibres for use in wide variety of applications [32]. The majority of these polymers are derived from polymeric substrates that are either natural or synthetic in origin. The conventional electrospinning method is depicted in Figure 4.2, which may be seen here. It is made up of a spinneret that has a needle made out of metal attached to it, an injectable pump, a source of high voltage, and a linked collection unit [33]. Although horizontal and vertical configurations are typically ynthesi, electrospinning in an upward direction has also been applied in some circumstances [34, 35].

Essentially, the mixture/blend, composite, sol-gel, or polymeric solution/melt relying on organic or synthetic origin is put inside the syringe, and then by using a syringe pump, it is compelled towards the tip of the needle, where it produces a circular droplet. This step is repeated until the mixture/blend, sol-gel, composite, or polymeric solution/melt is complete. A voltage ranging from 5 to 40 kV is applied to the mixture that is located on the needle, which causes the drop to expand into a conical shape that is known as the Taylor cone [36]. An electrified jet can form and flow it towards the oppositely charged carrying collector depending on the fluid viscosity, which should be sufficient to maintain stretching as well as whipping while forming the nano structured particles. During this excursion, the solvent is allowed to evaporate, and the jet then solidifies, resulting in the formation of microfibre webs on the collector. The jet is only stable up until the point where the needle points, at which time it begins to become unstable. Remarkably, this method gives the processor the ability to change the features and physical properties of the solution in order to regulate the morphology and structure of the nanofibres that are created. The influence of a variety of electrospinning process parameters for wastewater membrane fabrication has been the subject of numerous articles [36, 37]. In addition, the morphology and structure of the generated nanofibre are significantly impacted by the solvent as well as the co-solvent for fabricating membrane filters for the wastewater treatment process.

## 4.3 TYPES OF ELECTROSPUN MEMBRANE FABRICATION PROCESS

Electrospinning process can be subdivided into following categories based on membrane fabrication process used for wastewater treatment process (Figure 4.3).

### 4.3.1 FABRICATION OF TRI-AXIAL ELECTROSPUN FIBRES FOR WATER FILTRATION MEMBRANE

Tri-axial electrospinning is a novel technique that has been developed by a number of scientists as an alternative to coaxial electrospinning [38]. The technique involves the use of a syringe with three needles arranged in a circle. Three distinct types of liquids are pumped up through the nozzles by three different pumps in a sequential order. Researchers made multi-layered biodegradable nanofibres using tri-axial electrospinning as their construction technique [39]. Gelatin was used for both the core and the sheath, whereas PCL was used for the layer in the centre. The sheath had a thickness of 130 nm, the intermediate layer had a thickness of 240 nm, and the core layer had a thickness of 230 nm. Additionally, soluble and insoluble polymer-based solutions, which were previously impossible to electrospin utilising traditional

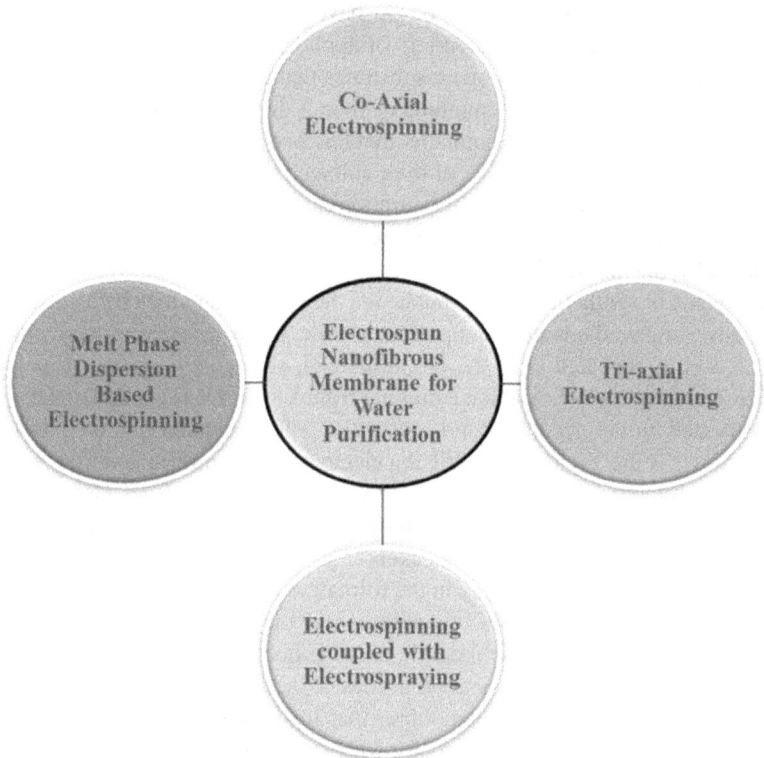

**FIGURE 4.3** Types of Electrospun Nanofibre Generation Process for Developing Membrane Filter for Water Purification Process.

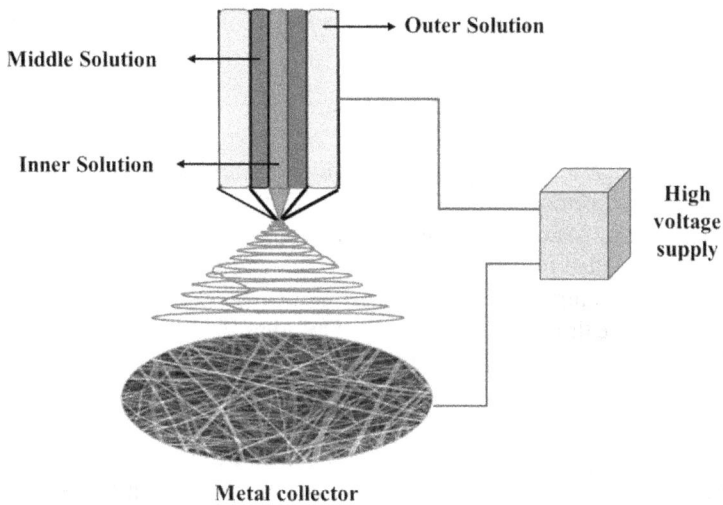

**FIGURE 4.4**    Basic Setup for Tri-Axial Electrospinning Process for Nanofibrous Membrane Fabrication for Water Purification.

co-axial electrospinning, are now able to be made by infusing a solvent along with the different solutions while maintaining an appropriate flow rate [39]. The included multilayers are capable of carrying out a wide range of activities and can incorporate the functional groups in order to ynthesiz the properties of the nanofibres to specific applications to trap pollutant particles from wastewater. Figure 4.4 illustrates the basic setup for the tri-axial electrospinning process.

## 4.3.2   FABRICATION OF CO-AXIAL ELECTROSPUN FIBRES FOR WATER FILTRATION MEMBRANE

Electrospinning in some kind of a coaxial direction allows for the production of empty or core-shell fibres that are easily capable of being functionalised metal oxide (MO)-based nanoparticles [40]. It is possible to ynthesize a wide number of components inside the core of the embedding substance in order to generate integrated multifunctional materials. These components include polysaccharides, oligosaccharides, nanoparticles, metallic salts, oils, proteins, and enzymes. The fabrication of nanofibres from substances that are difficult to electrospin in a typical manner or that are not suitable for electrospinning can be accomplished through the use of the coaxial electrospinning process [41]. It can be employed in situations that need the stability as well as the controlled release of very small molecules [42]. A variety of different functionalised complex geometries can be generated by carefully selecting the solvent and constituents that are employed in the process [43]. Utilising this method resulted in the production of collagen-r-poly("-caprolactone) structure composed of a core and a shell [44]. Poly-caprolactone (PCL) was first dissolved in ynthesizednio (TFE) as inner solution. Collagen that has been dissolved in TFE was placed in the outer solution [44].

### 4.3.3   Fabrication of Electrospun Fibres for Water Filtration Membrane by Coupling of Electrospinning with Electro-Spraying

The combination of electrospinning and electro-spraying is one method that may be employed for successful fabrication of an electrospun nanofibre-based membrane. In electro-spraying, the atomisation of a liquid solution is accomplished through the application of an electrical force. The charges that have collected over the surface of the needle cause a Coulomb's repulsion among themselves, which causes the doped mixture to be expelled from the tip of the needle. At some point in the future, it will be transformed into highly charged particles that have a higher potential electrically. The diameter of the droplet can be controlled by adjusting the feed velocity as well as the voltage settings. Electrospinning and electro-spraying, when combined, have been shown to be an ideal method for the fabrication of tunable photo electrodes that may be used in solar and fuel cells, as was established in earlier studies [45]. Electrospinning and electro-spraying both use the same electro-dynamic process to create their respective effects. Therefore, both of these processes can be combined into a single machine, which is referred to as a concurrent device of electrospinning and electro-spraying (SEE). This method was ynthesi to produce small $TiO_2$ particles and incorporate them inside the network of an electrospun non-woven nylon-6 membrane [46].

### 4.3.4   Fabrication of Electrospun Fibres for Water Filtration Membrane Using Melt Phase Splitting and Dispersion

For the production of considerable nanofibrous membranes, a method known as melt-phase separation and dispersed coating was adopted. This method is easy and straightforward and simple to comprehend [47]. Investigations on nano-composite photo-catalytic membranes have traditionally focused primarily on the membranes' individual photo-catalytic and filtration efficiencies, with just a small number of research looking at the membranes' synergistic output. This is owing to their inadequate mechanical resilience, lower grade reusability, and substantial porosity, all of which result in a weaker connection within the membrane matrix and the photo-catalyst. As a result, melt phase separation process has the ability to enhance the morphological and physical structure of the membrane, both of which are essential for the efficient operation of the membrane filtration process for wastewater treatment. The $TiO_2$ is combined with a synthetic organic polymer consisting of polyvinyl alcohol-co-ethylene during this procedure (PVA-co-PE). After that, during the process of melt casting, the solution is combined with cellulose acetate butyrate (CAB), which stands for cellulose acetate butyrate. The composite of $PVA$-co-$PE/CAB/TiO_2$ was drenched onto the surface of the polyproline substrate in order to form a nanofibre-based membrane for use in the purification and decomposition of methylene blue (MB) dye [47].

## 4.4   APPLICATION OF ELECTROSPUN NANOFIBRE AND ITS COMPOSITES FOR PRESSURE DRIVEN MEMBRANE FILTRATION UNIT FOR WATER TREATMENT

The design and construction processes for membranes that are predicated on electrospun nanofibre are straightforward. The inclusion of electrospun nanofibre in a membrane matrix has the potential to drastically reduce the total cost of an industrially ynthesi membrane when applied on a greater scale. Electrospun membranes are used in place of more traditional membranes since they provide an intriguing additional approach. The nanofibre has the potential to increase water flow while simultaneously reducing energy consumption. A great number of researchers in the past have ynthesi membranes made from electrospun nanofibres for the purpose of water purification, desalination, or as an adsorbing media for the removal of pollutants.

## 4.4.1 Reverse Osmosis (RO)

The technique known as reverse osmosis (RO), which, like nanofiltration (NF), is a process that is driven by pressure, is used to conduct desalination. RO membranes typically have pores that range in size from 0.1 to 1 nanometer in diameter. Membranes manufactured through the process of reverse osmosis (RO) are ynthesi in order to remove the tiniest particles, which include monovalent ions such as Cl-1 and Na+. In the past, RO membranes were created by fabricating them out of cellulose acetate (CA). A thin film composite RO membrane with a supplemental polymer film and an ultrathin preferred polyamide layer was ynthesi for the establishment of a desalination plant on a scale appropriate for commercial use [48]. It is necessary to improve the water permeability flux of the CA-containing RO membrane because it is currently poor. In order to address this issue, it is essential to accomplish the goal of ynthesize the porous structure by controlling the phase separator using an optimum mixture of solvents and non-solvents. In addition, the top and intermediate layers can be ynthesize during production by employing nano-composite materials or by combining the additives. When ynthesi for desalination with an increase in water flux, the newly ynthesized CA membranes have the potential to display a significant improvement in fouling resistance, chlorine resistance, antibacterial characteristics, mechanical properties, and thermal properties [48].

## 4.4.2 Nanofiltration (NF)

The nanofiltration (NF) membrane is able to separate particles varying in size from 100 to 1000 Da, and it is powered by pressure [49]. It is widely employed in the water treatment sector to eliminate colour, odour, ions, and trace amounts of organic pollutants. The Donnan repulsion and steric hindrance features of the NF systems allow for the effective elimination of multi-valent and divalent ions [49]. Nanofibres produced by electrospinning are used to cover the underlying surface of the thin film nanofiltration composite membrane. It is possible that the existence of electrospun nanofibre in the support layers will lead to an increase in flux as well as an improvement in the rejection rate. In order to create the NF membranes, a thin coating of polyamide was applied to the exterior of electrospun nanofibre. The NF membrane, in its acquired state, was shown to exhibit a higher rate of salt rejection in conjunction with an increased water flux. It was more effective in removing magnesium divalent cations and sulphate anions from the solution.

## 4.4.3 Ultrafiltration (UF)

The process known as ultrafiltration (UF) is one that employs a system that, when subjected to hydrostatic pressure, allows $H_2O$ molecules and smaller particles to pass through it. On the contrary, particles that are too large are not accepted. The pores of UF membranes commonly have a diameter ranging from 0.01 to 0.1 micrometres. Using a UF membrane, it is possible to separate viruses, colloids, and suspended particulate materials. Desalination and the purification of drinking water are therefore both applications that make use of ultra-filtration-based filters [50]. Electrospun nanofibres generally possess micro-pores. Because of this, they are ynthesi rather frequently in MF units. Ultrafiltration is a process that may be engineered into thin-film nanofibrous composites, also known as TFNC, membranes through the design and manufacturing processes. The TFNC membrane can be broken down into three separate layers. As the initial layer, one should employ both a hydrophilic coating that is non-porous and extremely thin or ultra-fine structured nano-fibres. Between the top layer – which is a nonwoven fabric – and the bottom layer, which is a micro-porous electrospun nanofibre, there is an intermediate level. Nanofibre-based membranes have a comparatively high porosity in comparison to traditional ultrafiltration (UF) membranes [50].

### 4.4.4  Microfiltration (MF)

The specific architecture and average pore size of electrospun nanofibres make them ideally suited for use in microfiltration (MF) units, regardless of where the nanofibres were produced. The use of MF makes it possible to separate pollutants with a greater molecular weight. In general, the MF device is able to filter out particles of a diameter ranging from 0.1 to 10 micrometres. When compared to traditional MF membranes, electrospun nanofibre-based MF membranes offer exceptional features such as enhanced porosity and the ability to modify the pore distribution. The efficacy of the membrane surface of the MF device is reliant on the size of the particle and the suspended substance. The diameter of the pores should be relatively smaller than the size of the suspended particles in the waste steam [51], as this would maximise the penetrability and selectivity of the pores. Figure 4.5 displays the basic principle of NF-, UF-, and MF-based membrane configuration for water treatment.

## 4.5  THE DRAWBACKS OF UTILISING ELECTROSPUN NANOFIBRE AND ITS COMPOSITES WASTEWATER FILTRATION/TREATMENT

A fundamental paradigm transition is currently taking place in the direction of the industrialisation aynthesizednionion of electrospynthesizeres for wastewater treatment and filtration approach. The new technological advancements in fibre gathering strategies, needle shapes, and higher output during production have proved successful in the generation of these fascinating nano-structured substances. In addition to the challenges that are inherently involved in increasing the production ynthesizeres, the quality ynthesizeres derived from the same source might vary greatly, whether they are derived from a synthetic or natural origin. However, the ultimate qualities are influenced by the environment in which the plant was grown. An additional environmental burden is imposed as a result of the very corrosive process that must be used to remove certain natural polymers. In order to achieve high levels of purity and quality in the natural and synthetic polymers that are produced on a large scale, it is

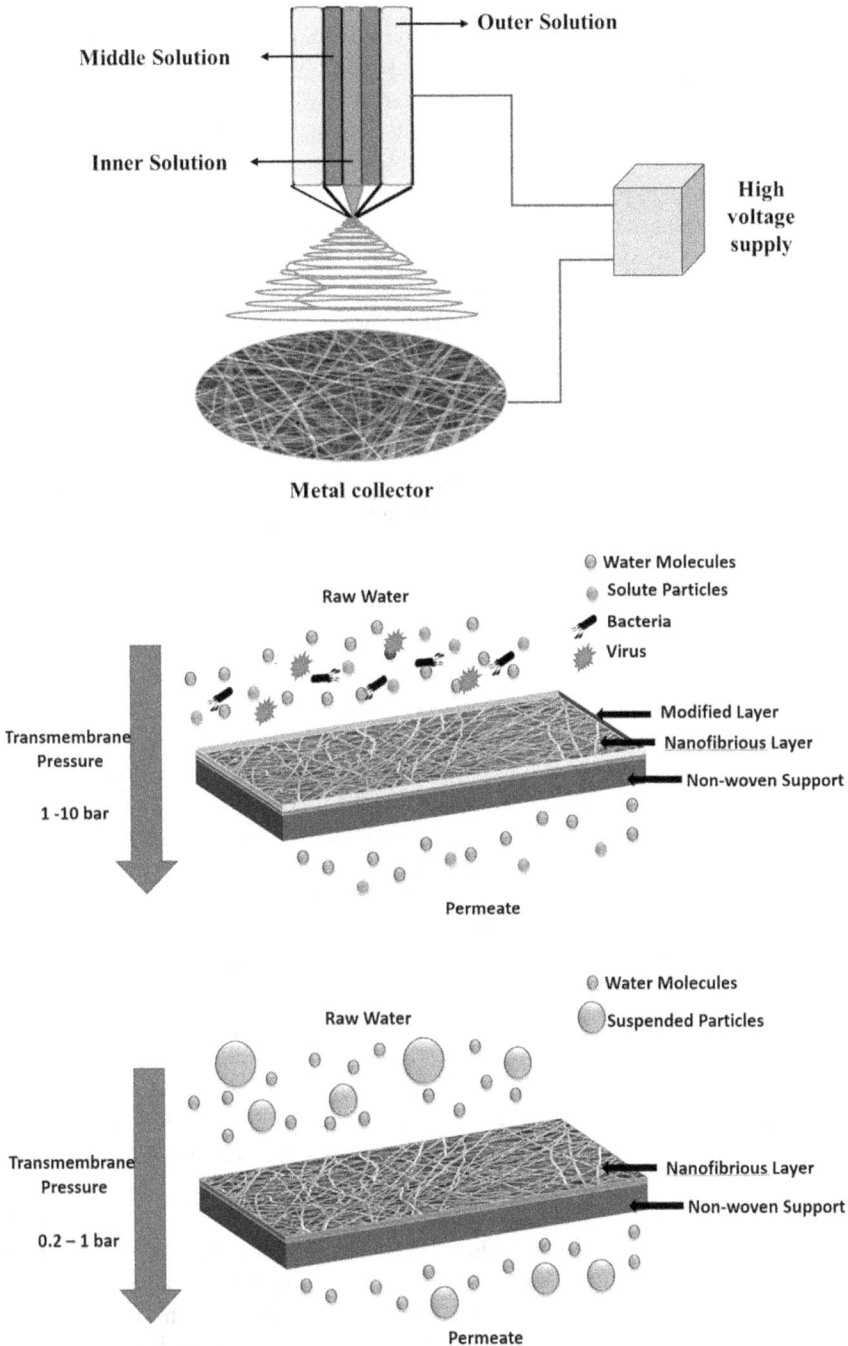

**FIGURE 4.5** Basic Principle of (a) NF, (b) UF, and (c) MF System for Water Treatment.

essential to conduct substantial research on genetically modified organisms and innovative engineering manufacturing techniques. Electrospun nanofibrous membranes are largely suited for use in membrane technologies that involve mild pressures. This is primarily the case because the mechanical strength that is necessary to withstand the pressures involved with water filtering is rather low (MF, UF, and NF). The use of electrospun nanofibres in water and wastewater treatment is still in its infant stages, despite the fact that research on these nanofibres is increasingly employing solvents that are less harmful to the environment. Their service life is too short, and as a result, they need to be replaced on a regular basis. This is due to the fact that the vast majority of these nanofibres are derived from natural sources – which means that they are often biodegradable – and that some of them are easily soluble in water. Because of this, the operating costs are increasing. In order for nanofibrous membranes to ynthesised in the water or wastewater purification or filtration process, it is essential for the membranes to ynthesizesed through physicochemical treatment. However, thynthesizednion must not compromise the membranes' distinctive properties. It takes a significant amount of knowledge and skill to accomplish the diffusion of nano-cluster metal oxide onto the electrospynthesizeres without aggregation, while also being able to ensure their accessibility and emission into the water introducing secondary pollutants.

## 4.6  CONCLUSION

It is completely obvious that electrospun nanofibre-based membranes are fairly fragile, unstable, and simple to break up into small pieces. This is mostly attributable to the huge pore size of the membranes. As a consequence of this, these kinds of membrane filtration units are typicalynthesised in the form of a mat and are simply submerged in contaminated water. With this procedure, on the other hand, additional processing is required in order to separate the mat from the filtered water. The installation of electrospun nanofibre-based membranes in treated wastewater might lead to secondary contamination if the membranes are not properly cleaned. As a result, it needs to be incorporated onto a platform that is sturdy and powerful in order to improve the membrane's mechanical strength and its level of customisability. In addition to this, it has the potential to make the recycling and maintenance of the membrane easier. According to the findings of our investigation, the majority of research studies suggested that the constructed electrospun nanofibrous membranes are instantaneously immersed in the contaminated water for photocatalytic degradation or oxidation of organic pollutants. However, the primary purpose of the membrane, which is selectiynthesizetion of pollutants with greater water flux, was ignored by these researchers. It wasn't until recently that significant work began to be done to develop leaching-prevention techniques for metal oxide-based photo-catalysts. There has only been a small amount of research that has been published on the topic of the significance and contribution made by the filtering of degraded pollutants from waste streams by these types of membranes. In addition, it is not understood how the process of filtering interacts with the photocatalytic degradation that occurs. In addition, there is a factual question that must be addressed through future research, and that question is whether or not filtration can increase the capacity for photocatalytic degradation. As a consequence of this, comprehensive investigations on this mechanism are necessary, which, if successful, will ideally drive future research and the effective application of MO-doped bi-functional nanofibrous polymeric membranes.

In the twenty-first century, the accumulation of waterborne pollutants can pose significant problems for individuals and communities, particularly in less developed nations. Even in modern times, drinking water in wealthy countries is not completely free of microorganisms. Those operations that involve the careless discharge of industrial effluents are the principal sources of the transfer of waterborne pathogens in addition to other inorganic and organic pollutants. Substantial efforts to address the problem have been prompted by the escalating need to find a solution to treat the wastewater. The application of nanomaterials and/or their combinations in the process of water purification is currently attracting a lot of attention in the scientific community. The interesting features of nanoparticles, along with their incorporation into already existing technologies, indicate a potentially bright future for tynthesizednion of wastewater treatment. Because of this, a large amount of research into the treatment of waste effluents including the integration of nanofibres and metal oxides with photocatalytic capabilities has been carried out by scientists and engineers. In this chapter, we discussed the current state of the art regarding the utilisation of electrospun nanofibre-based membranes, as well as probable forecasts regarding their future employment in the filtration of water. In this chapter, it is stated how the synthesis methodology centred on strengthening the structural and functional qualities of the membrane materials, along with their use for filtration of wastewater.

## ACKNOWLEDGMENTS

The authors are thankful for the funding provided by Malaysian Joint Research Scheme-ST 077–2022, Interdisciplinary Research IIRG003A-2022IISS and International Grant ICF 023–2022 and ICF 080–2021 under University of Malaya, Kuala Lumpur 50603, Malaysia. The authors would like to acknowledge Doctors Feed Ltd, Bangladesh, and University Utara Malaysia, Research opportunity with Grant Number RI 10010030/SO 21153 for their support in this work.

## REFERENCES

[1]    Vörösmarty, C. J., Hoekstra, A. Y., Bunn, S. E., Conway, D., and Gupta, J. 2015. Fresh water goes global. *Science*, 349, 478–479. DOI: 10.1126/science.aac6009.

[2]    Eliasson, J. 2015. The rising pressure of global water shortages. *Nature*, 517, 6. DOI: 10.1038/517006a.

[3]    Pendergast, M. M., and Hoek, E. M. V. 2011. A review of water treatment membrane nanotechnologies. *Energy Environmental Science*, 4, 1946–1971. DOI: 10.1039/c0ee00541j

[4]    Chong, M. N., Jin, B., C., Chow, W. K., and Saint, C. 2010. Recent developments in photocatalytic water treatment technology: A review. *Water Research*, 44, 2997–3027. DOI: 10.1016/j.watres.2010.02.039.

[5]    Padaki, M., Surya Murali, R., Abdullah, M. S., Misdan, N., Moslehyani, A., Kassim, M. A., et al. 2015. Membrane technology enhancement in oil–water separation. A review. *Desalination*, 357, 197–207. DOI: 10.1016/j.desal.2014.11.023.

[6]  Subramani, A., and Jacangelo, J. G. 2015. Emerging desalination technologies for water treatment: A critical review. *Water Research*, 75, 164–187. DOI: 10.1016/j.watres. 2015.02.032.

[7]  Ang, W. L., Mohammad, A. W., Hilal, N., and Leo, C. P. 2015. A review on the applicability of integrated/hybrid membrane processes in water treatment and desalination plants. *Desalination*, 363, 2–18. DOI: 10.1016/j.desal.2014.03.008.

[8]  Tran, N. H., Ngo, H. H., Urase, T., and Gind, K. Y.-H. 2015. A critical review on characterization strategies of organic matter for wastewater and water treatment processes. *Bioresouce Technology*, 193, 523–533. DOI: 10.1016/j.biortech.2015.06.091.

[9]  Lin, D.-J., Chang, H.-H., Chen, T.-C., Lee, Y.-C., and Cheng, L.-P. 2006. Formation of porous poly(vinylidene fluoride) membranes with symmetric or asymmetric morphology by immersion precipitation in the water/TEP/PVDF system. *European Polymer Journal*, 42, 1581–1594. DOI: 10.1016/j.eurpolymj.2006.01.027.

[10]  Schaep, J., and Vandecasteele, C. 2001. Evaluating the charge of nanofiltration membranes. *Journal of Membrane. Science*, 188, 129–136. DOI: 10.1016/S0376-7388(01) 00368-4

[11]  Van Der Bruggen, B., Vandecasteele, C., Van Gestel, T., Doyen, W., and Leysen, R. 2003. A review of pressure-driven membrane processes in wastewater treatment and drinking water production. *Environmental Progress*, 22, 46–56. DOI: 10.1002/ep.670220116.

[12]  Meng, F., Chae, S.-R., Drews, A., Kraume, M., Shin, H.-S., and Yang, F. 2009. Recent advances in membrane bioreactors (MBRs): Membrane fouling and membrane material. *Water Research*, 43, 1489–1512. DOI: 10.1016/j.watres.2008.12.044.

[13]  Kim, J., and Van der Bruggen, B. 2010. The use of nanoparticles in polymeric and ceramic membrane structures: Review of manufacturing procedures and performance improvement for water treatment. *Environmental Pollution*, 158, 2335–2349. DOI: 10.1016/j.envpol.2010.03.024.

[14]  Kang, G.-D., and Cao, Y.-M. 2014. Application and modification of poly(vinylidene fluoride) (PVDF) membranes—a review. *Journal of Membrane. Science*, 463, 145–165. DOI: 10.1016/j.memsci.2014.03.055.

[15]  Wang, C., de Bakker, J., Belt, C. K., Jha, A., Neelameggham, N. R., Pati, S., et al. 2013. *Energy Technology 2013: Carbon Dioxide Management and Other Technologies*. NJ: Wiley. DOI: 10.1002/9781118888735.

16]  Shao, P., and Huang, R. Y. M. 2007. Polymeric membrane pervaporation. *Journal of Membrane Science*, 287, 162–179. DOI: 10.1016/j.memsci.2006.10.043.

[17]  Mallada, R., and Menéndez, M. 2008. *Inorganic Membranes: Synthesis, Characterization and Applications*, Vol. 13. Elsevier, Amsterdam, p. 480.

[18]  Lv, D., Wang, R., Tang, G., Mou, Z., Lei, J., Han, J., et al. 2009. Ecofriendly electrospun membranes loaded with visible-light-responding nanoparticles for multifunctional usages: highly efficient air filtration, dye scavenging, and bactericidal activity. *ACS Applied Materials Interfaces*, 11, 12880–12889. http s://doi.org/10.1021/acsami.9b01508.

[19]  Ma, Z., and Ramakrishna, S. 2008. Electrospun regenerated cellulose nanofiber affinity membrane functionalized with protein A/G for IgG purification. *Journal of Membrane. Science*, 319(1–2), 23–28. https://doi.org/10.1016/j.memsci.2008.03.045.

[20]  Nasir, A. M., Jaafar, J., Aziz, F., Yusof, N., Norhayati, W., Salleh, W., Ismail, A. F., and Aziz, M. 2020. A review on floating nanocomposite photocatalyst: Fabrication and applications for wastewater treatment. *Journal of Water Processing Engineering*, 36, 101300. https://doi.org/10.1016/j.jwpe.2020.101300.

[21]  Yahya, N., Nasir, A. M., Daub, N. A., Aziz, F., Aizat, A., J. Jaafar, J., Lau, W. J., Yusof, N., Salleh, W. N. W., Ismail, A. F., and Aziz, M. 2020. Chapter 9—visible light–driven perovskite-based photocatalyst for wastewater treatment. In *Hand Book Smart Photocatalytic Mater.* Elsevier, pp. 265–302, https://doi.org/10.1016/B978-0-12-819051–7.00009–9.

[22] Leong, S., Razmjou, A., Wang, K., Hapgood, K., Zhang, X., and Wang, H. 2014. TiO$_2$ based photocatalytic membranes: A review. *Journal of Membrane Science*, 472, 167–184. https://doi.org/10.1016/j.memsci.2014.08.016.

[23] Samad, A., Furukawa, M., Katsumata, H., Suzuki, T., and Kaneco, S. 2016. Photocatalytic oxidation and simultaneous removal of arsenite with CuO/ZnO photocatalyst. *Journal of Photochemistry Photobiology A Chemistry*, 325, 97–103. https://doi.org/10.1016/j. jphotochem.2016.03.035.

[24] Alias, N. H. Jaafar, J., Samitsu, S., Ismail, A. F., Mohamed, M. A., Othman, M. H. D., Rahman, M. A., Othman, N. H., Nor, N. A. M., Yusof, N., and Aziz, F. 2020. Mechanistic insight of the formation of visible-light responsive nanosheet graphitic carbon nitride embedded polyacrylonitriynthesizeres for wastewater treatment. *Journal of Water Processing Engineering*, 33, 101015. https://doi.org/10.1016/j.jwpe.2019.101015.

[25] She, Q., Wang, R., Fane, A. G., and Tang, C. Y. 2016. Membrane fouling in osmotically driven membrane processes: A review. *Journal of Membrane. Science*, 499, 201–233. DOI: 10.1016/j.memsci.2015.10.040.

[26] Zhao, C., Xue, J., Ran, F., and Sun, S. 2013. Modification of polyethersulfone membranes–a review of methods. *Progress Materials Science*, 58, 76–150. DOI: 10.1016/j. pmatsci.2012.07.002

[27] Cuiming, W., Tongwen, X., and Weihua, Y. 2003. Fundamental studies of a new hybrid (inorganic–organic) positively charged membrane: membrane preparation and characterizations. *Journal of Membrane Science*, 216, 269–278. DOI: 10.1016/S0376-7388 (03)00082-6

[28] Duan, J., Pan, Y., Pacheco, F., Litwiller, E., Lai, Z., and Pinnau, I. 2015. High performance polyamide thin-film-nanocomposite reverse osmosis membranes containing hydrophobic zeolitic imidazolate framework-8. *Journal of Membrane Science*, 476, 303–310. DOI: 10.1016/j.memsci.2014.11.038

[29] Hong, J., and He, Y. 2014. Polyvinylidene fluoride ultrafiltration membrane blended with nano-ZnO particle for photo-catalysis self-cleaning. *Desalination*, 332, 67–75. DOI: 10.1016/j.desal.2013.10.026

[30] Balta, S., Sotto, A., Luis, P., Benea, L., Van der Bruggen, B., and Kim, J. 2012. A new outlook on membrane enhancement with nanoparticles: the alternative of ZnO. *Journal of Membrane. Science*, 389, 155–161. DOI: 10.1016/j.memsci.2011.10.025

[31] Zhao, L. H., Shen, L. G., He, Y. M., Hong, H. C., and Lin, H. J. 2015. Influence of membrane surface roughness on interfacial interactions with sludge flocs in a submerged membrane bioreactor. *Journal of Colloid Interface Science*, 446, 84–90. DOI: 10.1016/j.jcis.2015.01.009.

[32] Masilela, N., Kleyi P., Tshentu, Z., Priniotakis, G., Westbroek, P., and Nyokong, T. 2013. Photodynamic inactivation of Staphylococcus aureus using low symmetrically substituted phthalocyanines supported on a polystyrene polymer fiber. *Dyes and Pigments*, 96, 500–508. DOI: 10.1016/j.dyepig.2012.10.001

[33] Greiner, A., and Wendorff, J. H. 2007. Electrospinning: A fascinating method for the preparation of ultrathin fibers. *Angewandte Chemie International Edition*, 46, 5670–5703. DOI: 10.1002/anie.200604646.

[34] Huang, Z-M., Zhang, Y-Z., Kotaki, M., and Ramakrishna, S. 2003. A review on polymer nanofibers by electrospinning and their applications in nanocomposites. *Composites Science and Technology*, 63, 2223–2253. DOI: 10.1016/S0266–3538(03)00178-7.

[35] Abdel-Hady, F., Alzahrany, A., and Hamed, M. 2011. Experimental validation of upward electrospinning process. *ISRN Nanotechnology*, 2011, 851317/1–851317/14. DOI: 10.5402/2011/851317.

[36] Subbiah, T., Bhat, G., Tock, R., Parameswaran, S., and Ramkumar, S. 2005. Electrospinning of nanofibers. *Journal of Applied Polymer Science*, 96, 557–569. DOI: 10.1002/ app.21481.

[37] Bhardwaj, N., and Kundu, S. C. 2010. Electrospinning: A fascinating fiber fabrication technique. *Biotechnology Advances*, 28, 325–347. DOI: 10.1016/j.biotechadv.2010.01.004.

[38] Kim, W., and Kim, S. S. 2011. Synthesis of biodegradable triplelayered capsules using a triaxial electrospray method. *Polymer*, 52, 3325–3336. DOI: 10.1016/j.polymer.2011.05.033.

[39] Liu, W., Ni, C., Chase, D. B., and Rabolt, J. F. 2013. Preparation of multilayer biodegradable nanofibers by triaxial electrospinning. *ACS Macro Letters*, 2, 466–468. DOI: 10.1021/mz4000688.

[40] Tang, Y., Fu, S., Zhao, K., Teng, L., and Xie, G. 2016. Fabrication of TiO$_2$ micro-/nano-spheres embedded in nanofibers by coaxial electrospinning. *Materials Research Bulletin*, 78, 11–15. https://doi.org/10.1016/j.materresbull.2016.02.018.

[41] Tong, H. W., Zhang, X., and Wang, M. 2012. A new nanofiber fabrication technique based on coaxial electrospinning. *Materials Letters*, 66, 257–260. DOI: 10.1016/j.matlet.2011.08.095

[42] Chang, J-J., Lee, Y-H., Wu, M-H., Yang, M-C., and Chien, C. T. 2012. Preparation of electrospun alginate fibers with chitosan sheath. *Carbohydrate Polymers*, 87, 2357–2361. DOI: 10.1016/j.carbpol.2011.10.054.

[43] Li, D., and Xia, Y. 2004. Direct fabrication of composite and ceramic hollow nanofibers by electrospinning. *Nano Letters*, 4, 933–938. DOI: 10.1021/nl049590f.

[44] Zhang, Y., Huang, Z-M., Xu, X., Lim, C. T., and Ramakrishna, S. 2004. Preparation of core–shell structured PCL-rgelatin Bi-component nanofibers by coaxial electrospinning. Chemistry of *Materials*, 16, 3406–3409. DOI: 10.1021/cm049580f.

[45] Martin, S., and Castillo, J. L. 2018. Ten-fold reduction from the state-of-the-art platinum loading of electrodes prepared by electrospraying for high temperature proton exchange membrane fuel cells. *Electrochemical Communication*, 93, 57–61. https://doi.org/10.1016/j.elecom.2018.06.007.

[46] Pahasupanan, T., Suwannahong, K., Dechapanya, W., and Rangkupan, R. 2018. Fabrication and photocatalytic activity of TiO$_2$ composite membranes via simultaneous electrospinning and electrospraying process. *Journal of Environmental Science*, 72, 13–24. https://doi.org/10.1016/j.jes.2017.11.025.

[47] Ni, Y., Yan, K., Xu, F., Zhong, W., Zhao, Q., and Liu, K. 2019. Synergistic effect on TiO$_2$ doped poly (vinyl alcohol-co-ethylene) nanofibrous film for filtration and photocatalytic degradation of methylene blue. *Composites Communications*, 12, 112–116. https://doi.org/10.1016/j.coco.2019.01.007.

[48] Hailemariam, R. H., Woo, Y. C., Damtie, M. M., Kim, B. C., Park, K. D., and Choi, J. S. 2020. Reverse osmosis membrane fabrication and modification technologies and future trends: A review. *Advances in Colloid and Interface Science*, 276, 102100. https://doi.org/10.1016/j.cis.2019.102100

[49] Mahdavi, H., and Moslehi, M. 2016. A new thin film composite nanofiltration membrane based on PET nanofiber support and polyamide top layer: Preparation and characterization. *Journal of Polymer Research*, 23(12). https://doi.org/10.1007/s10965-016-1157-4

[50] Pereira, B. S., Moreti, L. O. R., Silva, M. F., Bergamasco, R., Piccioli, A. F. B., Garcia, E. E., Costa, W. V., Pineda, E. A. G., Oliveira, D. M. F., and Hechenleitner, A. A. W. 2017. Water permeability increase in ultrafiltration cellulose acetate membrane containing silver nanoparticles. *Material Research*, 20(Suppl 2), 887–891. https://doi.org/10.1590/1980-5373-mr-2016-1074.

[51] Arribas, P., García-Payo, M. C., Khayet, M., and Gil, L. 2019. Heat-treated optimized polysulfone electrospun nanofibrous membranes for high performance wastewater microfiltration. *Separation Purification Technology*, 226, 323–336. https://doi.org/10.1016/j.seppur.2019.05.097

# 5 Green Nanotechnology in Wastewater Treatment

*G. K. Prashanth, M. Mutthuraju, Manoj Gadewar,*
*Srilatha Rao, K. V. Yatish, Mithun Kumar Ghosh,*
*A. S. Sowmyashree and K. Shwetha*

## 5.1 INTRODUCTION

Water is an essential natural resource for human survival. The hydrosphere covers more than 70% of the earth's surface, yet only around 2.5% of it is accessible as clean water in vapour form, in atmosphere, as glaciers, groundwater, and rivers [1]. Of all water which is found in the sea, 97% is not accessible for drinking. The majority of the world's population is suffering from a shortage of fresh drinking water [2]. Demand for clean water has been steadily growing due to prolonged drought and population growth. Currently, wastewater is treated using traditional approaches like chemical, physical, and biological methods; however, advanced technology for wastewater treatment is urgently needed to meet the demand for fresh water. Nanotechnology is a favourable area that can be used to successfully treat wastewater. In many countries, the lack of proper wastewater treatment systems has led to increased pollution of the already scarce freshwater resources. Researchers have noted that this has led to advances in existing water treatment systems in both developing and developed countries [3–4]. There are various methods that can be used to treat polluted water, such as ultraviolet (UV) filtration, anion exchange, deionisation, liquid distillation, reverse osmosis, active carbon by adsorption, and ultrafiltration, but their performance is limited due to their complex operating process, low output/input ratio and high cost [5]. In light of these limitations, nanotechnology enters the picture, which not only provides a non-toxic environment but can also be used to conserve resources [6].

Nanomaterials/NPs considerably boost the adsorption capability of pollutant materials owing to their mass ratios to large surface area [7]. Inorganic components such as heavy metals, solvents, metallic ions; organic compounds like proteins, material of plants, food excreta; and pathogens such as coliform fecal streptococci, micro-organisms, and virus particles are frequent components of wastewater. Green nanomaterial-based tools (catalysts, adsorbents, etc.) can be used to develop environmentally friendly, efficient, and cost-effective wastewater management solutions. Nanomaterials are the particles or chemicals with dimensions ranging from 1 nm to 100 nm and nanotechnology is the process of synthesising and manipulating such materials. Compared to bulk materials, NPs have unique properties like high strength, conductivity, and better chemical reactivity. Green NPs are made from natural ingredients, which make them environmentally friendly and safe for humans

DOI: 10.1201/9781003342830-5

and the environment. Natural substances like vitamins, proteins, carbohydrates, plant extracts, peptides, and biopolymers act as reducing agents in NPs formation. The synthesis of green nanomaterials leads to a combination of green chemistry and nanotechnology. Metal oxide NPs, such as zinc oxide (ZnO), titanium dioxide ($TiO_2$), silver and iron oxide ($Fe_3O_4$) NPs, are often used as green NPs for water and wastewater treatment systems. For the manufacture of NPs from plant extracts, numerous elements such as nickel (Ni), aluminium (Al), manganese (Mn), cobalt (Co), palladium (Pd), zinc (Zn), copper (Cu), silver (Ag), platinum (Pt), and gold (Au) are commonly employed. The metal-based NPs from plant extracts have several applications such as catalysts/photocatalysts, magnetic property, deoxyribonucleic acid (DNA) binding, anti-carcinogenic activity, antibacterial activity, anti-tubercular activity, anti-fungal activity, heavy metal sensing, dye degradation, etc. [8–22].

## 5.2 NANOMATERIALS IN GREEN CONTEXT

Natural chemicals are used to make green nanomaterials, which make them environmentally friendly and safe for human health. Nanotechnology may be new but NPs are not new; these are abundant in nature, and many nanoscale things, such as proteins, enzymes, and DNA, are essential for life. The nanotechnology in green approach is the study of making nanomaterials from natural elements that are environmentally friendly [23]. Natural, biocompatible, and biodegradable nanomaterials are synthesised using green methods. Vitamins, proteins, carbohydrates, plant extracts, peptides, and biopolymers are among the naturally occurring compounds that serve as acceptable reducing agents for the creation of green nanomaterials. Nanotechnology in a green approach is grouping of nanotechnology and green chemistry that can be used to produce a long-term water and wastewater treatment system. Wastewater treatment by nanotechnology is depicted in Figure 5.1.

## 5.3 WHY PLANT EXTRACTS ARE MORE PROMISING THAN ORGANISMS

Green NPs are synthesised by using several living organisms like bacteria, fungi, leaf, root, bark, fruit, stem, flower, etc. Plants have been identified to be the most suitable species for the commercial production of green NPs due to the fact that the maintenance of bio-organisms (bacteria, fungi, etc.) is very costly. Plants make green NPs more reliable than microorganisms, and plant-based nanomaterials are more diverse in shape and size than the nanomaterials obtained from other green sources. Scanning electron microscopy (SEM), transmission electron microscopy (TEM), and nuclear power microscopy are some of the advanced microscopic techniques used in the morphological analysis of NPs during their synthesis.

## 5.4 TYPES OF NANOMATERIALS USED IN WASTEWATER TREATMENT

Metal-based nanomaterials, nanomembranes, nanocatalysts, metal-oxide nanomaterials, carbonaceous materials, magnetic NPs, and nano-biopolymers are some of

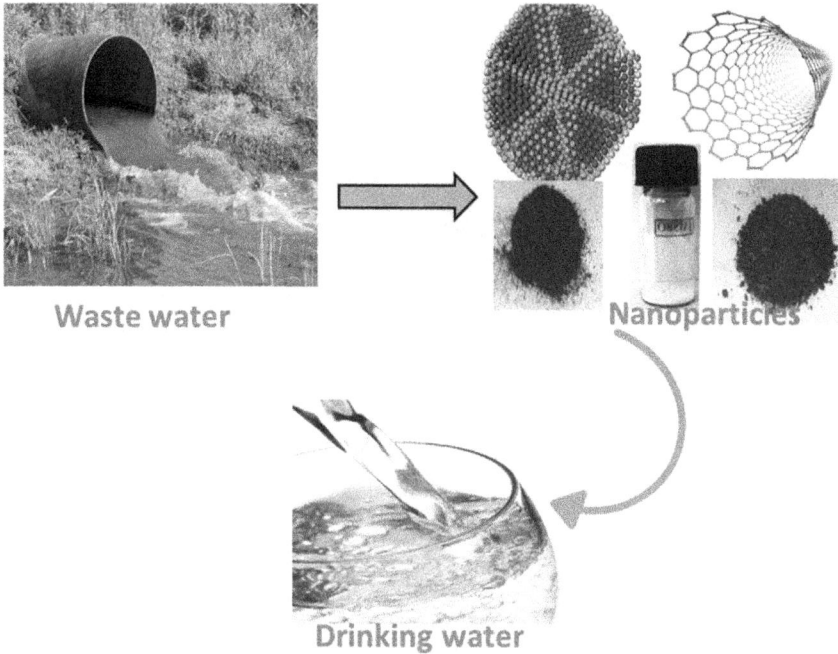

**FIGURE 5.1** Wastewater Treatment by Nanotechnology.

the nanomaterials used to treat water contaminated by radionuclides, toxic organic pathogens, toxic heavy metals, and chemicals.

### 5.4.1 METAL-BASED NANOMATERIALS

Metal NPs have a vast surface area, which allows them to adsorb tiny organic molecules. The adsorption of these molecules on solid surfaces has a variety of environmental and bioanalytical applications [24]. Metal NPs have become the subject of intense research and development in recent years due to their unique properties, such as enhanced catalytic properties, surface absorption properties, and much more reactivity. Several scientific studies have shown that nanomaterials have the ability to remove a wide range of toxic spectrum from water, indicating that they may be effective in treating wastewater. Silver, gold, and iron are the most examined metal-based nanomaterials in this field.

Due to its excellent physical chemical properties (strong electrical and thermal conductivity) as well as biological ability (antibacterial properties), the silver metal has been well known for its exceptional properties since the Roman era [25–29]. AgNP's (silver NPs) ability to disinfect drinking water when combined with filtration processes has been extensively studied. AgNPs are well known for their antimicrobial activity. Therefore, AgNPs are used to remove harmful microorganisms from wastewater. Previous research has indicated three probable pathways for AgNP antibacterial activity. First, AgNPs can attach to cellular membranes, disrupting cell permeability and respiratory processes, leading to cell death [30–32]. Second, reactive

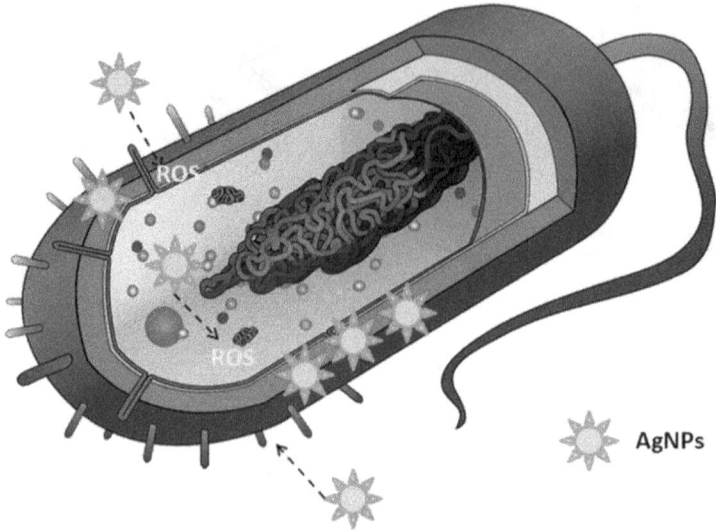

**FIGURE 5.2**   The Possible Mechanism of Anti-Bacterial Activity by AgNPs.

oxygen species (ROS) formed on NPs surfaces produce oxidative stress, which leads to significant DNA damage [33]. Third, the silver ions produced by AgNPs cause adenosine triposphate (ATP) synthesis and DNA replication to be disrupted [34, 35] Figure 5.2.

Currently, nanoscale zero valent iron (nZVI) is gaining attention to purify the wastewater [36–47]. Specific NPs were demonstrated to be very reactive in comparison to granular iron, which is commonly used in reactive barriers and in situ wastewater treatment due to their relatively tiny size and huge surface area. In anaerobic conditions, nZVI (Fe°) is oxidised by H⁺ or water and formation $H_2$ and $Fe^{2+}$ **(Equation I-II)**. $Fe^{2+}$ is further oxidised to form $Fe^{3+}$ which produces $Fe(OH)_3$. $Fe(OH)_3$ can remove the contaminants like chromium (VI) [48]. Moreover, nZVI has the ability to degrade the organic molecules from wastewater in the presence of oxygen. ZVI reacts with oxygen and produces hydrogen peroxide ($H_2O_2$) **(Equation III)**. $H_2O_2$ is reduced by ZVI and forms $H_2O$ **(Equation IV)**. Moreover, $H_2O_2$ reacts with $Fe^{2+}$ and forms highly reactive species hydroxyl radicals (.OH) which possesses strong oxidising ability **(Equation VI)** [49].

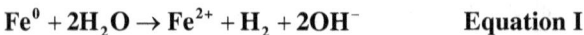

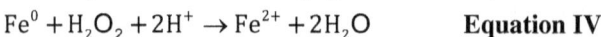

$$Fe^0 + 2H_2O \rightarrow Fe^{2+} + H_2 + 2OH^- \qquad \textbf{Equation I}$$

$$Fe^0 + 2H^+ \rightarrow Fe^{2+} + H_2 \qquad \textbf{Equation II}$$

$$Fe^0 + O_2 + 2H^+ \rightarrow Fe^{2+} + H_2O_2 \qquad \textbf{Equation III}$$

$$Fe^0 + H_2O_2 + 2H^+ \rightarrow Fe^{2+} + 2H_2O \qquad \textbf{Equation IV}$$

$$Fe^{2+} + H_2O_2 \rightarrow Fe^{3+} + {}^\circ OH + OH^- \qquad \textbf{Equation V}$$

Scheme 1: Possible reaction mechanism of wastewater treatment by nanoscale zero valent iron

## 5.4.2 NANOCATALYST

Owing to their unusual large surface to mass ratio, unique characteristic shape, and capacity to boost the surface catalytic activity, nanocatalysts are widely used in water treatment [50]. They help to degrade environmental pollutants like nitro aromatics [51], pesticides [52], polychlorinated biphenyls [53], halogenated herbicides [54], and azo-dyes [55].

## 5.4.3 MAGNETIC NPs

MNPs (Magnetic NPs) are one of the types of NPs that are made up of roughly 70 different elements, including cobalt, nickel, and iron. Depending on their biological compatibility, magnetic sensitivity, synthesis mode, and characterisation methods, these materials are operated by magnetic bar and deployed for relevant purposes. MNPs have always been proved to be effective in both the environmental and biomedical field such as diagnostic applications [56–58], cancer therapy [59], targeted drug delivery [60–63], Alzheimer's treatment [64, 65], and antimicrobial activities [66]. MNPs are widely recognised as a versatile technique for removing a variety of contaminants from soil, the air, and water as shown in Figure 5.3 [67–69].

## 5.4.4 NANO MEMBRANES

Membranes made of nanofibres that can remove micron-sized particles from the aqueous phase at a high rate are being used in a pre-treatment step before ultrafiltration or reverse osmosis [70]. The insertion of nanomaterials in polymeric or inorganic membranes, known as composite nanomembranes, has been addressed in numerous studies involving membrane nanotechnology. Metal-oxide NPs such as $TiO_2$, silica,

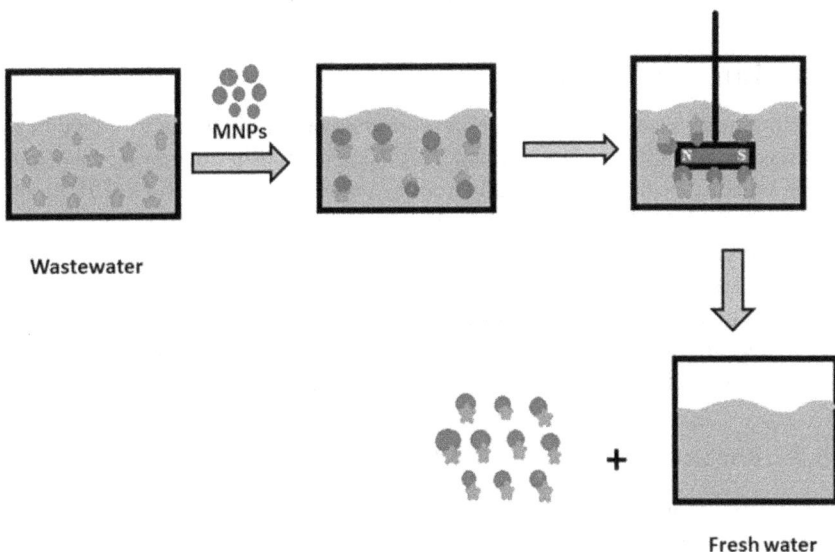

**FIGURE 5.3** Wastewater Treatment by MNPs.

**TABLE 5.1**
**Types of NPs and their applications in wastewater treatment**

| NPs | Application Wastewater | Reference |
|---|---|---|
| Metal-based nanomaterials | Organic dye degradation, antibacterial activity | **25–29** |
| Silver and gold NPs | Heavy metal ions detection, $H_2O_2$ detection | **30–32** |
| Nanocatalyst | Photocatalytic activity, Degradation of pesticides | **50–55** |
| Zeolite | Ultrafiltration membranes to increase their hydrophilicity and water permeability | **71–76** |
| nZVI | Halogenated organic compounds and $H_2O_2$ degradation | **48–49** |
| Magnetic NPs | Removal of a variety of contaminants from soil, air, and water | **67–69** |

alumina, and zeolite have been introduced into polymeric ultrafiltration membranes to increase their hydrophilicity and water permeability [71–76]. A few case studies of NPs and their applications in wastewater treatment are provided in Table 5.1.

## 5.5 WASTEWATER TREATMENT BY GREEN NPS

In wastewater treatment systems, green nanomaterials such as colloids, nanowires, quantum dots, nanotubes, films, and particles are used. Nanomaterials are distinguished by their small size and large surface area, as well as their superior absorption properties and strong reducing capacity, which contribute to the removal of toxins from wastewater [77]. Four types of green nanomaterial application techniques for water and wastewater treatment have been discussed here: filtration technique, pollutant adsorption by NPs, organic compounds removal by NPs, and heavy metal and ions removal by NPs.

### 5.5.1 FILTRATION TECHNIQUE

Nano-fibres are used in nano-membranes to remove micro-circulated contaminants from reservoirs and to reduce fouling [78]. Nanocomposite membranes are multifunctional advance membranes made from a mixture of nanomaterials [79]. In ultrafiltration or reverse osmosis, nanomembranes are used as a pre-treatment measure. $TiO_2$, zeolite, Alumina, and silica NPs were introduced into ultrafiltration polymeric membranes to improve water permeability, fouling resistance, and surface hydrophilicity. For decomposing/breakdown of chlorinated chemicals, inorganic membrane $TiO_2$ NPs were used [80]. The production of bacterial biofilms in polymeric membranes containing silver NPs was significantly inhibited.

### 5.5.2 POLLUTANT ADSORPTION BY NPS

Nanomaterials can adsorb contaminants from water, with the adsorbing material's surface containing certain functional groups that interact with the contaminant's ionic groups. Nanomaterials or nanostructured components have very high specific surface areas, possibly associated with functional groups, creating low expansion distances, specific pore sizes, and surface chemical absorbers [81]. Carbon nanotubes, manganese oxide, ferric oxides, ZnO, TiO$_2$, and magnesium oxide are some of the most-often employed NPs for pollutant adsorption [82].

## 5.5.3 ORGANIC COMPOUNDS REMOVAL BY NPS

Alkyl chloride, benzenes derivatives, and polychlorinated biphenyls (PCBs) are the examples of natural contaminants that are both hydrophilic and hydrophobic [83, 84]. In addition to cerium oxide (CeO$_2$), TiO$_2$ NPs are used as catalysts for the complete degradation of natural pollutants as well as ozonation processes. [85, 86]. Organic dyes such as methyl red (MR), methylene blue (MB), naphthyl orange (NO), and methyl orange (MO) contaminate the drinking water. Nanocatalysts have the ability to reduce organic pigments from contaminated water in presence of sunlight. Many researchers are working in this field. Organic pollutants such as azo dyes, pesticides, and PCBs can be decomposed by bimetallic nanocatalysts containing zerovalence metals and semiconductor materials [87]. SA-AgNPs (*Sonchus arvensis* silver NPs) can degrade MB within 1 hour in presence of sunlight [88]. Cu based zirconium dioxide (ZrO$_2$) nanocatalysts have photocatalytic activity against the pollutant dye NO. Within 85 minutes, 80.2% NO was degraded by nanocatalyst [89]. Y$_2$ZnO$_4$ is a promising nanocatalyst. Y$_2$ZnO$_4$ degraded MO within 65 minutes. The authors have

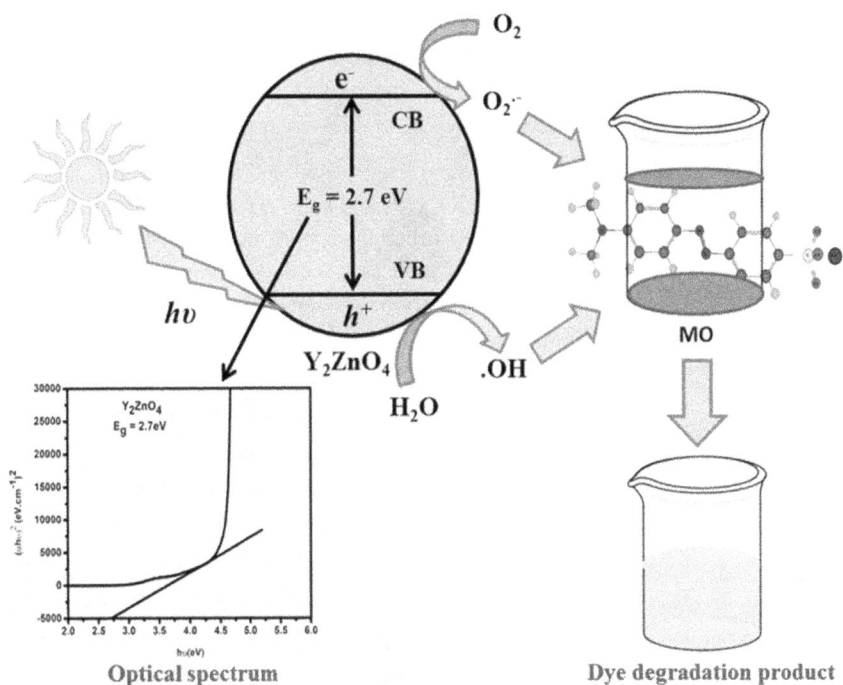

**FIGURE 5.4**   Methyl Orange Degradation by Y$_2$ZnO$_4$ in Presence of Sunlight.
Source: Reprinted with permission from Ref. [90]. Further permissions related to the material excerpted should be directed to the ACS)

shown the mechanism for MO degradation in the presence of the nanocatalyst and sunlight as shown in Figure 5.4 [90].

Generally, NPs adsorb photon from sunlight electron (e$^-$) and hole (h$^+$) pair (**Equation VI**). The electron (e$^-$) and hole (h$^+$) are called conduction band (CB) and valence band (VB) respectively. The e- reacts with the oxygen and produces the $O_2^-$ (superoxide) radicals (**Equation VII**). The h$^+$ reacts with water molecules to generate hydroxyl anion (OH$^-$) and H$^+$ (**Equation VIII**). Another molecule of water reacts with h$^+$ and forms reactive species. OH (**Equation IX**). The organic dye (OD) is activated by sunlight and forms OD* (**Equation X**) and OD$^+$ (**Equation XI**). OD$^+$ reacts with active species. OH and $O_2^-$ and decomposes the dye (**Equation XII and XIII**).

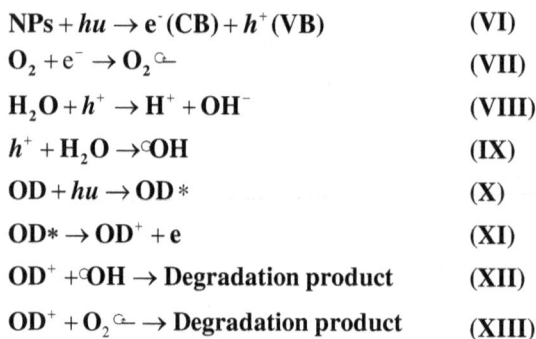

$$\mathbf{NPs} + hu \rightarrow \mathbf{e^-(CB)} + h^+(\mathbf{VB}) \qquad \text{(VI)}$$

$$\mathbf{O_2} + e^- \rightarrow \mathbf{O_2^{\cdot-}} \qquad \text{(VII)}$$

$$\mathbf{H_2O} + h^+ \rightarrow \mathbf{H^+} + \mathbf{OH^-} \qquad \text{(VIII)}$$

$$h^+ + \mathbf{H_2O} \rightarrow \text{°OH} \qquad \text{(IX)}$$

$$\mathbf{OD} + hu \rightarrow \mathbf{OD*} \qquad \text{(X)}$$

$$\mathbf{OD*} \rightarrow \mathbf{OD^+} + \mathbf{e} \qquad \text{(XI)}$$

$$\mathbf{OD^+} + \text{°OH} \rightarrow \textbf{Degradation product} \qquad \text{(XII)}$$

$$\mathbf{OD^+} + \mathbf{O_2^{\cdot-}} \rightarrow \textbf{Degradation product} \qquad \text{(XIII)}$$

Scheme 2: Possible reaction mechanism of photocatalytic reaction

### 5.5.4   HEAVY METAL AND IONS REMOVAL BY NPS

Heavy metal salt was effectively removed from wastewater using metal-based NPs. Arsenic was adsorbed from water by $TiO_2$ NPs and nano-sized magnetite. In the presence of sunshine, Hg$^{2+}$ and Pb$^{2+}$ were adsorbed from wastewater by carbon nanotubes (CNTs)/$Fe_3O_4$ NPs [91]. Heavy metals such as arsenic can be removed using iron oxide NPs ($Fe_3O_4$ and $Fe_2O_3$) [92]. SA-AgNPs can remove the heavy metal ions Fe$^{3+}$ and Hg$^{2+}$ and its possible mechanism is shown in Figure 5.5 [88]. At a pH range of 3 to 7, the polymer-grafted ferric oxide ($Fe_2O_3$) nanocomposite successfully removed di-valent metal ions like cobalt, copper, and nickel. 3D nanostructures of $CeO_2$ were used as good adsorbents for the removal of heavy metals such as Cr and As [93]. Few examples of green NPs and their applications are presented in Table 5.2.

### 5.6   FUTURE PLAN

Nanotechnology has emerged as one of the most significant water treatment methods over the last few decades. Concerns about declining freshwater supply and increasing demand over time have led to a huge search for innovative green wastewater treatment solutions. There is a debate among experts as to whether nanotechnology can be considered a green strategy in this area [94]. NPs will provide effective alternative methods for wastewater treatment in future. One such technique is membrane bioreactors [95]. In the near future new and unique varieties of NPs will emerge that will be lighter, stronger, more efficient, better defined, and easier to integrate into existing systems [96, 97].

**TABLE 5.2**
**Green NPs and Wastewater Treatment**

| Entry | Name of NPs | Applications | Ref |
|---|---|---|---|
| 1 | JC-CuNPs | Degradation of MB from wastewater | 8 |
| 2 | AH-CuNPs | Removal of congo red and bacterial from wastewater | 10 |
| 3 | BP-AgNPs | Degradation of $H_2O_2$, MB and rhodamine-B from wastewater | 15 |
| 4 | AC-AgNPs | Degradation of $H_2O_2$ and MB from wastewater | 18 |
| 5 | JA@ AgNPs | Removal of $H_2O_2$ from wastewater | 20 |
| 6 | Ferragels | Removal of Cr and Pb from wastewater | 37 |
| 7 | Fe doped $TiO_2$ NPs | Degradation of diazinon in aqueous solution | 52 |
| 8 | $CuFe_2O_4$/MWCNTs | Degradation of diethyl phthalate in aqueous solution | 53 |
| 9 | $Fe_3O_4$@ His@ Cu | Degradation of azo dyes in aqueous solution | 55 |
| 10 | $Fe_3O_4$@ $SiO_2$-EDTA | Pb (II) and Cu (II) removal from aqueous | 69 |
| 11 | Graphene membranes | Reverse osmosis | 76 |
| 12 | $TiO_2$ nanowire | Photocatalytic oxidation of humic acid in wastewater | 80 |
| 13 | $TiO_2$ NPs | Degradation of organic contaminants from wastewater | 84 |
| 14 | $CeO_2$, $TiO_2$ NPs | Degradation of natural pollutants as well as ozonation process | 85–86 |
| 15 | SA-AgNPs | Removal of $Hg^{2+}$, $Fe^{2+}$ and dye MB from wastewater | 88 |
| 16 | Cu based $ZrO_2$ | Removal of NO from wastewater | 89 |
| 17 | $Y_2ZnO_4$ | Degradation of MO from wastewater | 90 |
| 18 | MPTS-CNTs/$Fe_3O_4$ | Adsorbed $Hg^{2+}$ and $Pb^{2+}$ from wastewater | 91 |
| 19 | $Fe_3O_4$ and $Fe_2O_3$ NPs | Removal of heavy metal ions from wastewater | 92 |
| 20 | $CeO_2$ NPs | Removal of Cr and As from wastewater | 93 |

## 5.7 SUMMARY

When the concentration of one or more compounds rises to a point where it causes problems for living creatures and the environment, the substances are referred to as contaminants and the situation is referred to as water contamination. Effective and innovative water treatment technologies are urgently needed to ensure safe drinking water by removing microproliferators and improving industrial production treatment processes through the use of flexible water treatment systems. Reservoirs can naturally clean a certain amount of pollution through bacterial breakdown or innocuous dispersal, but in many instances the concentration of pollution is so high that it cannot be cleaned naturally. Nanotechnology-based water treatment systems promise innovative therapeutic capabilities that allow cost-effective use of unusual water sources to expand water supply. Nanomaterials have been proposed as effective, low cost, and environmentally friendly treatment materials. The sensing and detecting process, the pollution prevention mechanism, and treatment and remediation process

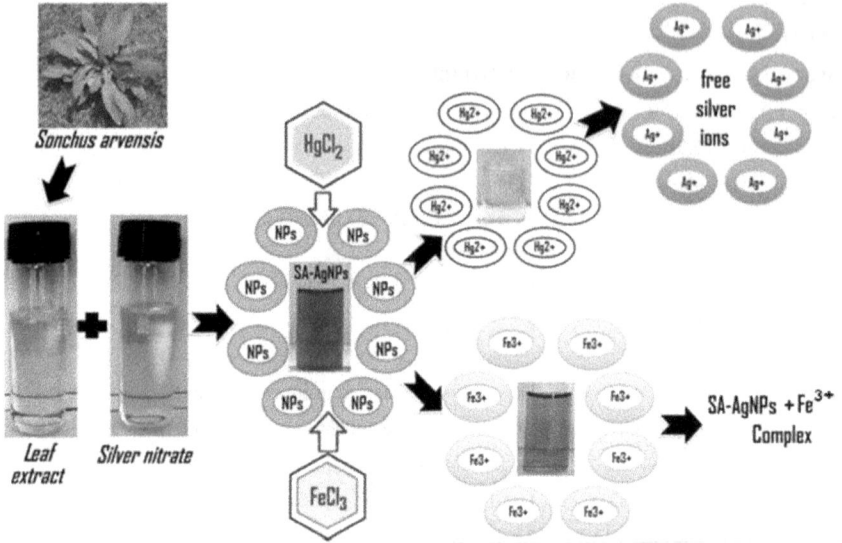

**FIGURE 5.5**   $Fe^{3+}$ and $Hg^{2+}$ removal by SA-AgNPs [88].

are the three key categories of green nanomaterial applications in wastewater treatment. Biological materials, organic compounds, chemicals, heavy metal ions such as nickel, arsenic, mercury, arsenic, lead, cadmium and copper could all be traced and eliminated using nanopowder, NPs, and nanomembranes. Green NPs are being used to treat wastewater and are helping to ensure water safety.

## REFERENCES

[1]   Black, M., 2016. *The atlas of water: Mapping the world's most critical resource.* University of California Press. ISBN: 9780520292031.

[2]   Amin, M. T., Alazba, A. A. and Manzoor, U., 2014. A review of removal of pollutants from water/wastewater using different types of nanomaterials. *Advances in Materials Science and Engineering, 2014*, pp. 2–9.

[3]   Zhao, B., Su, Y., He, S., Zhong, M. and Cui, G., 2016. Evolution and comparative assessment of ambient air quality standards in China. *Journal of Integrative Environmental Sciences, 13*(2–4), pp. 85–102.

[4]   Jang, Y. T., Choi, C. H., Ju, B. K., Ahn, J. H. and Lee, Y. H., 2003. Suppression of leakage current via formation of a sidewall protector in the microgated carbon nanotube emitter. *Nanotechnology, 14*(5), p. 497.

[5]   Patanjali, P., Singh, R., Kumar, A. and Chaudhary, P., 2019. Nanotechnology for water treatment: A green approach. In *Green synthesis, characterization and applications of NPs* (pp. 485–512). Elsevier.

[6]   Gupta, N., Pant, P., Gupta, C., Goel, P., Jain, A., Anand, S. and Pundir, A., 2018. Engineered magnetic NPs as efficient sorbents for wastewater treatment: a review. *Materials Research Innovations, 22*(7), pp. 434–450.

[7]   DeFriend, K. A., Wiesner, M. R. and Barron, A. R., 2003. Alumina and aluminate ultrafiltration membranes derived from alumina NPs. *Journal of Membrane Science, 224*(1–2), pp. 11–28.

[8] Ghosh, M. K., Sahu, S., Gupta, I. and Ghorai, T. K., 2020. Green synthesis of copper NPs from an extract of Jatropha curcas leaves: Characterization, optical properties, CT-DNA binding and photocatalytic activity. *RSC Advances*, *10*(37), pp. 22027–22035.

[9] Prashanth, G. K., Prashanth, P. A., Bora, U., et al., 2015. In vitro antibacterial and cytotoxicity studies of ZnO nanopowders prepared by combustion assisted facile green synthesis. *Karbala International Journal of Modern Science*, *1*, pp. 67–77.

[10] Chandraker, S. K., Lal, M., Ghosh, M. K., Tiwari, V., Ghorai, T. K. and Shukla, R., 2020. Green synthesis of copper NPs using leaf extract of Ageratum houstonianum Mill. and study of their photocatalytic and antibacterial activities. *Nano Express*, *1*(1), p. 010033.

[11] Prashanth, G. K., Prashanth, P. A., Nagabhushana, B. M., Ananda, S., Krishnaiah, G. M., Nagendra, H. G., Sathyananda, H. M., Rajendra Singh, C., Yogisha, S., Anand, S. and Tejabhiram, Y., 2018. Comparison of anticancer activity of biocompatible ZnO NPs prepared by solution combustion synthesis using aqueous leaf extracts of Abutilon indicum, Melia azedarach and Indigofera tinctoria as biofuels. *Artificial Cells, Nanomedicine, and Biotechnology*, *46*(5) (August), pp. 968–979.

[12] Chandraker, S. K., Ghosh, M. K., Lal, M. and Shukla, R., 2021. A review on plant-mediated synthesis of silver NPs, their characterization and applications. *Nano Express*, *2*(2), p. 022008.

[13] Krishna, P. G., Ananthaswamy, P. P., Trivedi, P., Chaturvedi, V., Bhangi Mutta, N., Sannaiah, A., Erra, A. and Yadavalli, T., 2017. Antitubercular activity of ZnO NPs prepared by solution combustion synthesis using lemon juice as bio-fuel. *Materials Science and Engineering C*, *75*, pp. 1026–1033.

[14] Prashanth, G. K., Prashanth, P. A., Meghana Ramani, A. S., Nagabhushana, B. M., Krishnaiah, G. M., Nagendra, H. G., Sathyananda, H. M., Mutthuraju, M. and Rajendra Singh, C., 2019. Comparison of antimicrobial, antioxidant and anticancer activities of ZnO NPs prepared by lemon juice and citric acid fueled solution combustion synthesis. *BioNanoScience*, *9*, pp. 799–812.

[15] Chandraker, S. K., Lal, M., Dhruve, P., Singh, R. P. and Shukla, R., 2021. Cytotoxic, antimitotic, DNA binding, photocatalytic, H2O2 sensing, and antioxidant properties of biofabricated silver NPs using leaf extract of Bryophyllum pinnatum (Lam.) Oken. *Frontiers in Molecular Biosciences*, p. 465.

[16] Prashanth, G. K., Prashanth, P. A., Padam Singh, B. M., Nagabhushana, C., Shivakumara, K. G. M., Nagendra, H. G., Sathyananda, H. M. and Chaturvedi, V., 2020. Effect of doping (with cobalt or nickel) and UV exposure on the antibacterial, anticancer, and ROS generation activities of zinc oxide NPs. *Journal of Asian Ceramic Societies*, *8*(4), pp. 1175–1187.

[17] Sathyananda, H. M., Prashanth, P. A., Prashanth, G. K., Nagabhushana, B. M., Shivakuma-ra, C., Boselin Prabhu, S. R. and Nagendra, H. G., 2022. Evaluation of antimycobacterial, antioxidant, and anticancer activities of CuO NPs through cobalt doping. *Applied Nano Science- Springer*, *12*, pp. 79–86.

[18] Prashanth, G. K., Sathyananda, H. M., Prashanth, P. A., et al., 2022 Controlled synthesis of Ag/CuO nanocomposites: evaluation of their antimycobacterial, antioxidant, and anticancer activities. *Applied Physics A*, *128*, p. 614. https://doi.org/10.1007/s00339-022-05748-x.

[19] Krishna, P. G., Prashanth, P. A., Bora, U., Gadewar, M. and Bhangi Mutta, N., 2017. In vitro antibacterial and anticancer studies of ZnO NPs prepared by sugar fueled combustion synthesis. *Advanced Materials Letters*, *8*, pp. 24–29.

[20] Chandraker, S. K., Lal, M., Kumar, A. and Shukla, R., 2021. Justicia adhatoda L. mediated green synthesis of silver NPs and assessment of their antioxidant, hydrogen peroxide sensing and optical properties. *Materials Technology*, pp. 1–11.

[21] Prashanth, G. K., Mutthuraju, M., Manoj, M., Gadewar, P. P. A., Srilatha Rao, B. P., Yatish, K. V. and Nagendra, H. G. Photocatalytic activity induced by metal nanoparticles synthesized by sustainable approaches- A comprehensive review. *Photocatalysis and Photochemistry*. doi: 10.3389/fchem.2022.917831, 2022.

[22] Tamang, A. M., Singh, N., Chandraker, S. K. and Ghosh, M. K., 2021. Solvent impregnated resin a potential alternative material for separation dyes, metal and phenolic compounds: A review. *Current Research in Green and Sustainable Chemistry*, p. 100232.

[23] Ahmed, S., Saifullah, Ahmad, M., Swami, B. L. and Ikram, S., 2016. Green synthesis of silver NPs using Azadirachta indica aqueous leaf extract. *Journal of Radiation Research and Applied Sciences*, 9(1), pp. 1–7.

[24] Gehrke, I., Geiser, A. and Somborn-Schulz, A., 2015. Innovations in nanotechnology for water treatment. *Nanotechnology, Science and Applications*, 8, p. 1.

[25] Tang, C., Hu, D., Cao, Q., Yan, W. and Xing, B., 2017. Silver NPs-loaded activated carbon fibers using chitosan as binding agent: Preparation, mechanism, and their antibacterial activity. *Applied Surface Science*, 394, pp. 457–465.

[26] Gurunathan, S., Park, J. H., Han, J. W. and Kim, J. H., 2015. Comparative assessment of the apoptotic potential of silver NPs synthesized by Bacillus tequilensis and Calocybe indica in MDA-MB-231 human breast cancer cells: targeting p53 for anticancer therapy. *International Journal of Nanomedicine*, 10, p. 4203.

[27] Furlan, P. Y., Fisher, A. J., Melcer, M. E., Furlan, A. Y. and Warren, J. B., 2017. Preparing and testing a magnetic antimicrobial silver nanocomposite for water disinfection to gain experience at the nanochemistry–microbiology interface. *Journal of Chemical Education*, 94(4), pp. 488–493.

[28] Roy, A., Butola, B. S. and Joshi, M., 2017. Synthesis, characterization and antibacterial properties of novel nano-silver loaded acid activated montmorillonite. *Applied Clay Science*, 146, pp. 278–285.

[29] Joshi, M., Purwar, R., Udakhe, J. S. and Sreedevi, R., Indian Institute of Technology Delhi, 2015. *Antimicrobial nanocomposite compositions, fibers and films*. U.S. Patent, pp. 9, 192, 625.

[30] Park, M. V., Neigh, A. M., Vermeulen, J. P., de la Fonteyne, L. J., Verharen, H. W., Briedé, J. J., van Loveren, H. and de Jong, W. H., 2011. The effect of particle size on the cytotoxicity, inflammation, developmental toxicity and genotoxicity of silver NPs. *Biomaterials*, 32(36), pp. 9810–9817.

[31] Foldbjerg, R., Dang, D. A. and Autrup, H., 2011. Cytotoxicity and genotoxicity of silver NPs in the human lung cancer cell line, A549. *Archives of Toxicology*, 85(7), pp. 743–750.

[32] Park, E. J., Yi, J., Kim, Y., Choi, K. and Park, K., 2010. Silver NPs induce cytotoxicity by a Trojan-horse type mechanism. *Toxicology in Vitro*, 24(3), pp. 872–878.

[33] Feng, Q. L., Wu, J., Chen, G. Q., Cui, F. Z., Kim, T. N. and Kim, J. O., 2000. A mechanistic study of the antibacterial effect of silver ions on Escherichia coli and Staphylococcus aureus. *Journal of Biomedical Materials Research*, 52(4), pp. 662–668.

[34] Kittler, S., Greulich, C., Diendorf, J., Koller, M. and Epple, M., 2010. Toxicity of silver NPs increases during storage because of slow dissolution under release of silver ions. *Chemistry of Materials*, 22(16), pp. 4548–4554.

[35] Liu, J., Sonshine, D. A., Shervani, S. and Hurt, R. H., 2010. Controlled release of biologically active silver from nanosilver surfaces. *ACS Nano*, 4(11), pp. 6903–6913.

[36] Zhang, W. X., 2003. Nanoscale iron particles for environmental remediation: an overview. *Journal of Nanoparticle Research*, 5(3), pp. 323–332.

[37] Ponder, S. M., Darab, J. G. and Mallouk, T. E., 2000. Remediation of Cr (VI) and Pb (II) aqueous solutions using supported, nanoscale zero-valent iron. *Environmental Science & Technology*, 34(12), pp. 2564–2569.

[38] Adeleye, A. S., Conway, J. R., Garner, K., Huang, Y., Su, Y. and Keller, A. A., 2016. Engineered nanomaterials for water treatment and remediation: Costs, benefits, and applicability. *Chemical Engineering Journal*, *286*, pp. 640–662.

[39] Madhura, L., Singh, S., Kanchi, S., Sabela, M. and Bisetty, K., 2019. Nanotechnology-based water quality management for wastewater treatment. *Environmental Chemistry Letters*, *17*(1), pp. 65–121.

[40] Westerhoff, P., Alvarez, P., Li, Q., Gardea-Torresdey, J. and Zimmerman, J., 2016. Overcoming implementation barriers for nanotechnology in drinking water treatment. *Environmental Science: Nano*, *3*(6), pp. 1241–1253.

[41] Phillips, D. H., Nooten, T. V., Bastiaens, L., Russell, M. I., Dickson, K., Plant, S., Ahad, J. M. E., Newton, T., Elliot, T. and Kalin, R. M., 2010. Ten year performance evaluation of a field-scale zero-valent iron permeable reactive barrier installed to remediate trichloroethene contaminated groundwater. *Environmental Science & Technology*, *44*(10), pp. 3861–3869.

[42] Groundwater, R. T. C., 2010. Ten year performance evaluation of a field-scale zero-valent iron permeable reactive barrier installed to remediate trichloroethene contaminated groundwater. *Environmental Science & Technology*, *44*(10), pp. 3861–3869.

[43] Wilkin, R. T., Acree, S. D., Ross, R. R., Puls, R. W., Lee, T. R. and Woods, L. L., 2014. Fifteen-year assessment of a permeable reactive barrier for treatment of chromate and trichloroethylene in groundwater. *Science of the Total Environment*, *468*, pp. 186–194.

[44] Li, S., Wang, W., Liang, F. and Zhang, W. X., 2017. Heavy metal removal using nanoscale zero-valent iron (nZVI): Theory and application. *Journal of Hazardous Materials*, *322*, pp. 163–171.

[45] Stefaniuk, M., Oleszczuk, P. and Ok, Y. S., 2016. Review on nano zerovalent iron (nZVI): from synthesis to environmental applications. *Chemical Engineering Journal*, *287*, pp. 618–632.

[46] Li, S., Wang, W., Liu, Y. and Zhang, W. X., 2014. Zero-valent iron NPs (nZVI) for the treatment of smelting wastewater: A pilot-scale demonstration. *Chemical Engineering Journal*, *254*, pp. 115–123.

[47] Thakor, A. S., Jokerst, J., Zavaleta, C., Massoud, T. F. and Gambhir, S. S., 2011. Gold NPs: A revival in precious metal administration to patients. *Nano Letters*, *11*(10), pp. 4029–4036.

[48] Wang, Y., Fang, Z., Kang, Y. and Tsang, E. P., 2014. Immobilization and phytotoxicity of chromium in contaminated soil remediated by CMC-stabilized nZVI. *Journal of Hazardous Materials*, *275*, pp. 230–237.

[49] Crane, R. A. and Scott, T. B., 2012. Nanoscale zero-valent iron: future prospects for an emerging water treatment technology. *Journal of Hazardous Materials*, *211*, pp. 112–125.

[50] Sushma, D. and Richa, S., 2015. Use of NPs in water treatment: A review. *International Journal of Environmental Research*, *4*(10), pp. 103–106.

[51] Zhao, X., Lv, L., Pan, B., Zhang, W., Zhang, S. and Zhang, Q., 2011. Polymer-supported nanocomposites for environmental application: A review. *Chemical Engineering Journal*, *170*(2–3), pp. 381–394.

[52] Tabasideh, S., Maleki, A., Shahmoradi, B., Ghahremani, E. and McKay, G., 2017. Sonophotocatalytic degradation of diazinon in aqueous solution using iron-doped $TiO_2$ NPs. *Separation and Purification Technology*, *189*, pp. 186–192.

[53] Zhang, X., Feng, M., Qu, R., Liu, H., Wang, L. and Wang, Z., 2016. Catalytic degradation of diethyl phthalate in aqueous solution by persulfate activated with nano-scaled magnetic CuFe2O4/MWCNTs. *Chemical Engineering Journal*, *301*, pp. 1–11.

[54] Sangami, S. and Manu, B., 2018. Catalytic efficiency of laterite-based FeNPs for the mineralization of mixture of herbicides in water. *Environmental Technology*, *40*(20), pp. 2671–2683.

[55]    Kurtan, U., Amir, M., Baykal, A., Sözeri, H. and Toprak, M. S., 2016. Magnetically recyclable Fe3O4@ His@ Cu nanocatalyst for degradation of azo dyes. *Journal of Nanoscience and Nanotechnology*, *16*(3), pp. 2548–2556.

[56]    Stone, R. C., Fellows, B. D., Qi, B., Trebatoski, D., Jenkins, B., Raval, Y., Tzeng, T. R., Bruce, T. F., McNealy, T., Austin, M. J. and Monson, T. C., 2015. Highly stable multi-anchored magnetic NPs for optical imaging within biofilms. *Journal of Colloid and Interface Science*, *459*, pp. 175–182.

[57]    Raval, Y. S., Stone, R., Fellows, B., Qi, B., Huang, G., Mefford, O. T. and Tzeng, T. R. J., 2015. Synthesis and application of glycoconjugate-functionalized magnetic NPs as potent anti-adhesion agents for reducing enterotoxigenic Escherichia coli infections. *Nanoscale*, *7*(18), pp. 8326–8331.

[58]    Ishay, R. B., Israel, L. L., Eitan, E. L., Partouche, D. M. and Lellouche, J. P., 2016. Maghemite-human serum albumin hybrid NPs: Towards a theranostic system with high MRI r 2* relaxivity. *Journal of Materials Chemistry B*, *4*(21), pp. 3801–3814.

[59]    Huo, Y., Yu, J. and Gao, S., 2022. Magnetic nanoparticle-based cancer therapy. In *Synthesis and biomedical applications of magnetic nanomaterials* (pp. 261–290). EDP Sciences.

[60]    Ramaswamy, B., Kulkarni, S. D., Villar, P. S., Smith, R. S., Eberly, C., Araneda, R. C., Depireux, D. A. and Shapiro, B., 2015. Movement of magnetic NPs in brain tissue: mechanisms and impact on normal neuronal function. *Nanomedicine: Nanotechnology, Biology and Medicine*, *11*(7), pp. 1821–1829.

[61]    Kavre, I., Kostevc, G., Kralj, S., Vilfan, A. and Babič, D., 2014. Fabrication of magneto-responsive microgears based on magnetic nanoparticle embedded PDMS. *RSC Advances*, *4*(72), pp. 38316–38322.

[62]    D'Agata, F., Ruffinatti, F. A., Boschi, S., Stura, I., Rainero, I., Abollino, O., Cavalli, R. and Guiot, C., 2017. Magnetic NPs in the central nervous system: targeting principles, applications and safety issues. *Molecules*, *23*(1), p. 9.

[63]    Mohammed, L., Gomaa, H. G., Ragab, D. and Zhu, J., 2017. Magnetic NPs for environmental and biomedical applications: A review. *Particuology*, *30*, pp. 1–14.

[64]    McNamara, K. and Tofail, S. A., 2017. NPs in biomedical applications. *Advances in Physics: X*, *2*(1), pp. 54–88.

[65]    Mourtas, S., Lazar, A. N., Markoutsa, E., Duyckaerts, C. and Antimisiaris, S. G., 2014. Multifunctional nanoliposomes with curcumin–lipid derivative and brain targeting functionality with potential applications for Alzheimer disease. *European Journal of Medicinal Chemistry*, *80*, pp. 175–183.

[66]    Martinez-Gutierrez, F., Boegli, L., Agostinho, A., Sánchez, E. M., Bach, H., Ruiz, F. and James, G., 2013. Anti-biofilm activity of silver NPs against different microorganisms. *Biofouling*, *29*(6), pp. 651–660.

[67]    Huang, Y. and Keller, A. A., 2015. EDTA functionalized magnetic nanoparticle sorbents for cadmium and lead contaminated water treatment. *Water Research*, *80*, pp. 159–168.

[68]    Clark, K. K. and Keller, A. A., 2012. Adsorption of perchlorate and other oxyanions onto magnetic permanently confined micelle arrays (Mag-PCMAs). *Water Research*, *46*(3), pp. 635–644.

[69]    Liu, Y., Fu, R., Sun, Y., Zhou, X., Baig, S. A. and Xu, X., 2016. Multifunctional nanocomposites Fe3O4@ SiO2-EDTA for Pb (II) and Cu (II) removal from aqueous solutions. *Applied Surface Science*, *369*, pp. 267–276.

[70]    Giwa, A., Akther, N., Dufour, V. and Hasan, S. W., 2016. A critical review on recent polymeric and nano-enhanced membranes for reverse osmosis. *Rsc Advances*, *6*(10), pp. 8134–8163.

[71]    Akther, N., Daer, S. and Hasan, S. W., 2018. Effect of flow rate, draw solution concentration and temperature on the performance of TFC FO membrane, and the potential

use of RO reject brine as a draw solution in FO–RO hybrid systems. *Desalin. Water Treat*, *136*, pp. 65–71.

[72] Lim, S., Han, D. S., Pathak, N., Akther, N., Phuntsho, S., Park, H. and Shon, H. K., 2019. Efficient fouling control using outer-selective hollow fiber thin-film composite membranes for osmotic membrane bioreactor applications. *Bioresource Technology*, *282*, pp. 9–17.

[73] Fujiwara, M., 2017. Water desalination using visible light by disperse red 1 modified PTFE membrane. *Desalination*, *404*, pp. 79–86.

[74] Xiao, K., Kong, X. Y., Zhang, Z., Xie, G., Wen, L. and Jiang, L., 2016. Construction and application of photoresponsive smart nanochannels. *Journal of Photochemistry and Photobiology C: Photochemistry Reviews*, *26*, pp. 31–47.

[75] Pendergast, M. M. and Hoek, E. M. V., 2011. A review of water treatment membrane nanotechnologies. *Energy & Environmental Science*, *4*, pp. 1946–1971.

[76] Zhang, Z., Zhang, F., Liu, Z., Cheng, G., Wang, X. and Ding, J., 2018. Molecular dynamics study on the Reverse Osmosis using multilayer porous graphene membranes. *Nanomaterials*, *8*(10), p. 805.

[77] Theron, J., Walker, J. A. and Cloete, T. E., 2008. Nanotechnology and water treatment: applications and emerging opportunities. *Critical Reviews in Microbiology*, *34*(1), pp. 43–69.

[78] Srinivasan, S., Harrington, G. W., Xagoraraki, I. and Goel, R., 2008. Factors affecting bulk to total bacteria ratio in drinking water distribution systems. *Water Research*, *42*(13), pp. 3393–3404.

[79] Dotzauer, D. M., Dai, J., Sun, L. and Bruening, M. L., 2006. Catalytic membranes prepared using layer-by-layer adsorption of polyelectrolyte/metal nanoparticle films in porous supports. *Nano Letters*, *6*(10), pp. 2268–2272.

[80] Zhang, X., Du, A. J., Lee, P., Sun, D. D. and Leckie, J. O., 2008. TiO$_2$ nanowire membrane for concurrent filtration and photocatalytic oxidation of humic acid in water. *Journal of Membrane Science*, *313*(1–2), pp. 44–51.

[81] Gubin, S. P., Koksharov, Y. A., Khomutov, G. B. and Yurkov, G. Y., 2005. Magnetic NPs: preparation, structure and properties. *Russian Chemical Reviews*, *74*(6), p. 489.

[82] Sadegh, H., Shahryari-ghoshekandi, R. and Kazemi, M., 2014. Study in synthesis and characterization of carbon nanotubes decorated by magnetic iron oxide NPs. *International Nano Letters*, *4*(4), pp. 129–135.

[83] Kabra, K., Chaudhary, R. and Sawhney, R. L., 2004. Treatment of hazardous organic and inorganic compounds through aqueous-phase photocatalysis: A review. *Industrial & Engineering Chemistry Research*, *43*(24), pp. 7683–7696.

[84] Gaya, U. I. and Abdullah, A. H., 2008. Heterogeneous photocatalytic degradation of organic contaminants over titanium dioxide: A review of fundamentals, progress and problems. *Journal of Photochemistry and Photobiology C: Photochemistry Reviews*, *9*(1), pp. 1–12.

[85] Nawrocki, J. and Kasprzyk-Hordern, B., 2010. The efficiency and mechanisms of catalytic ozonation. *Applied Catalysis B: Environmental*, *99*(1–2), pp. 27–42.

[86] Kasprzyk-Hordern, B., Ziółek, M. and Nawrocki, J., 2003. Catalytic ozonation and methods of enhancing molecular ozone reactions in water treatment. *Applied Catalysis B: Environmental*, *46*(4), pp. 639–669.

[87] Orge, C. A., Órfão, J. J., Pereira, M. F., de Farias, A. M. D., Neto, R. C. R. and Fraga, M. A., 2011. Ozonation of model organic compounds catalysed by nanostructured cerium oxides. *Applied Catalysis B: Environmental*, *103*(1–2), pp. 190–199.

[88] Chandraker, S. K., Ghosh, M. K., Lal, M., Ghorai, T. K. and Shukla, R., 2019. Colorimetric sensing of Fe3+ and Hg2+ and photocatalytic activity of green synthesized silver NPs from the leaf extract of Sonchus arvensis L. *New Journal of Chemistry*, *43*(46), pp. 18175–18183.

[89]  Ghosh, M. K., Sahu, S. K., Sahoo, D. and Ghorai, T. K., 2020. Green synthesis, char-
      acterization and photocatalytic study of Cu based ZrO. *Journal of the Indian Chemical
      Society*, *97*(9b), pp. 1507–1513.
[90]  Ghosh, M. K., Jain, K., Khan, S., Das, K. and Ghorai, T. K., 2020. New dual-functional
      and reusable bimetallic Y2ZnO4 nanocatalyst for organic transformation under micro-
      wave/green conditions. *ACS Omega*, *5*(10), pp. 4973–4981.
[91]  Zhang, C., Sui, J., Li, J., Tang, Y. and Cai, W., 2012. Efficient removal of heavy metal
      ions by thiol-functionalized superparamagnetic carbon nanotubes. *Chemical Engineer-
      ing Journal*, *210*, pp. 45–52.
[92]  Takafuji, M., Ide, S., Ihara, H. and Xu, Z., 2004. Preparation of poly (1-vinylimidazole)-
      grafted magnetic NPs and their application for removal of metal ions. *Chemistry of
      Materials*, *16*(10), pp. 1977–1983.
[93]  Zhong, L. S., Hu, J. S., Cao, A. M., Liu, Q., Song, W. G. and Wan, L. J., 2007. 3D flow-
      erlike ceria micro/nanocomposite structure and its application for water treatment and
      CO removal. *Chemistry of Materials*, *19*(7), pp. 1648–1655.
[94]  Patanjali, P., Singh, R., Kumar, A. and Chaudhary, P., 2019. Nanotechnology for water
      treatment: A green approach. In *Green synthesis, characterization and applications of
      NPs* (pp. 485–512). Elsevier.
[95]  Neoh, C. H., Noor, Z. Z., Mutamim, N. S. A. and Lim, C. K., 2016. Green technology
      in wastewater treatment technologies: integration of membrane bioreactor with various
      wastewater treatment systems. *Chemical Engineering Journal*, *283*, pp. 582–594.
[96]  Berekaa, M. M., 2016. Nanotechnology in wastewater treatment; influence of nano-
      materials on microbial systems. *International Journal of Current Microbiology and
      Applied Sciences*, *5*(1), pp. 713–726.
[97]  Singh, P., Yadav, S. K. and Kuddus, M., 2020. Green nanomaterials for wastewater
      treatment. In *Green Nanomaterials* (pp. 227–242). Springer.

# 6 Integration of Microbial Treatment for Advanced Biological Treatment of Wastewater

*R. Reevenishaa Ravi Chandran,*
*Zaira Zaman Chowdhury, Masud Rana,*
*Ahmed Elsayid Ali, Rahman Faizur Rafique,*
*Md. Mahfujur Rahaman and*
*Karthickeyan Viswanathan*

## 6.1 INTRODUCTION

Recently, the impressive improvement in the industrial sectors has developed the human life dramatically. However, this significant improvement created new challenges via producing numerous amounts of highly toxic effluents such as aromatic hydrocarbons, ammoniacal nitrogen, dyes, pharmaceutical wastes, total dissolved solids (TDS), herbicides, pesticides, and heavy metals [1]. Although the improvements in the technological sectors have offered new solutions for industrial wastes treatment, the novel approach of treating industrial effluent revolve around hybrid techniques as highly efficient and economical treatment techniques [2]. Thus, this book chapter cover the several types of effluents produced by industrial sectors and methodologies adopted for eliminating the presence of these toxic effluent in water. Additionally, the novel and eco-friendly techniques used for removing toxic elements from wastewater using microbial species are illustrated. Wastewater treatment technologies are illustrated by Figure 6.1.

Generally, the development of countries is represented by their industrial and economic growth. Therefore, the industrial sectors exhibited a rapid growth in several countries. Typically, the industrialisation activities concentrate in areas close to urban centres. As a direct consequence of this, extraordinarily large stresses are placed on the ability of the ecosystem in these particular regions. Indeed, industrial sites consume the accessible natural around the area such as lakes, rivers, ponds, and seaside waters which cause environmental risks. These risks are formed by the discharging of the industrial effluents directly into the natural water resources. Basically, wastewater and industrial effluent can be defined as highly toxic by-products produced from human activities that are related to raw-material dispensation and manufacturing. The sources of wastewater may include wastewater produced by washing, cooking,

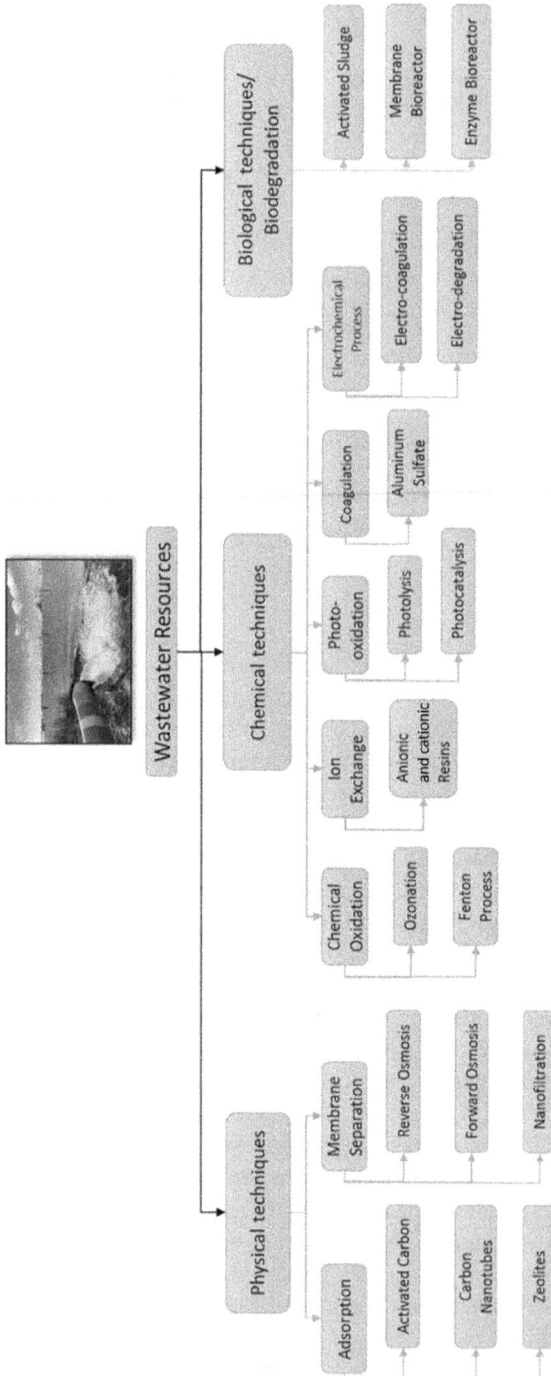

**FIGURE 6.1** Different Types of Water Treatment Technology.

reaction by-products, and separation. Conventional water treatment approaches exhibited an extremely low efficiency. Thus, utilising these approaches causes an increase in the accumulation of the toxic effluent. Nevertheless, the development of the new treatment approaches has overcome some of the challenges related to the conventional techniques. In general, the goal of the water treatment techniques is to eliminate the existence of the toxic materials to improve the water quality. Moreover, treating contaminated water will also eliminate the presence of substances that can influence the quality of water used for industrial applications. Moreover, treating water will reduce the impact of chemical oxygen demand (COD) and biological oxygen demand (BOD) in the environment. Above all, utilising novel water treatment techniques will increase the quality of the water to meet the global standard for potable water. In order to accomplish the objectives of the water treatment project, the water treatment plant must be designed and built in a kind of route that takes into consideration particular aspects of the influent that needs to be controlled in a more effective manner. The environmental protection agencies have set strict rules and standards to meet the best quality of water for human use. In fact, the rules and standards of the agencies are influenced by the different types of toxic chemicals that contaminate the water. From the economical perspective, treating water in high quality will reduce the cost of the risks caused by consuming contaminated water [3, 4]. The constituents of wastewater are shown by Figure 6.2.

As mentioned before, the primary source of water contaminates are unconstrained industries that discharge toxic effluents into the environment. Consuming contaminated water can lead to serious illnesses such as disorders of immunological function, lung and respiratory diseases, and certain other cancer-causing conditions [5, 6]. Around 60% of the wastewater is said to pose possible dangers to both the environment and public health as stated by the State Inspectorate for the Protection of the Environment (PIOS). Water treatment management makes use of vulnerability assessment such as workplace hazards, exposure assessment, and risk management models in order to guarantee that the water that is treated will have a high quality [7]. In addition to harmful chemicals, industrial effluent may also contain bacteria such as *E. coli*, *Salmonella*, and other similar organisms [8]. These bacteria are responsible for typhoid fever, cholera, as well as other allergic manifestations [8]. Thus, removal of some microbes from wastewater is also necessary, which has been effectively done using nano and micro-porous carbon along with some metal oxide nanoparticles (MO). However, some microbial species can be integrated also in

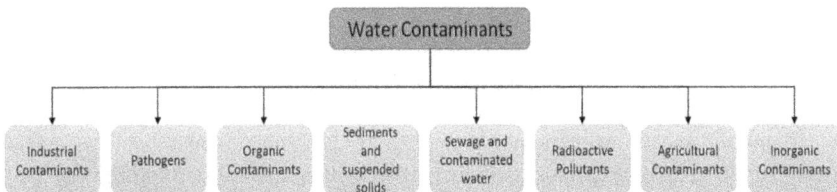

**FIGURE 6.2** Constituents of Wastewater.

wastewater treatment plants for biological pre-treatment and purification of water up to a certain extent. Myeloid leukemia is a disease primarily affecting workers who are overexposed to formaldehyde without taking a significant number of precautions in the past [9]. Asthma-like syndrome is a non-allergic respiratory disorder which is medically comparable to asthma but does not entail chronic inflammation of the airways or excessive reactivity [10].

Microalgae culture is an economically effective and environmentally benign technology for treating wastewater since it combines tertiary bio-based treatment with the generation of beneficial biomass [11, 12]. Because of the capability of microalgae to utilise inorganic phosphorus and nitrogen for its proliferation, microalgae cultures provide a comprehensive alternative to tertiary level and quaternary level processes. This makes microalgae cultures an attractive option. They are also known for their ability to remove heavy metals and certain hazardous organic compounds, allowing them to contribute to the prevention of secondary water contamination [13, 14].

As a substitute to traditional technologies such as those relying on activated sludge, the recovery of nutrients from effluent can be accomplished with the help of microalgae and bacteria. The natural sources of bacterial consortium predominate in the reactors because sterile circumstances are not conceivable in wastewater treatment frameworks. The presence of this instinctively occurring microbial consortium is a feature of the wastewater constituents, environmental factors, reactor configuration, and operational parameters [15]. Bacteria consume the existing organic wastes and produce carbon dioxide as a by-product. The photosynthetic process carried out by the microalgae results in the generation of carbon dioxide and carbohydrates. Carbohydrates are required for the formation of biomass, while oxygen is the ultimate electrophile for aerobic cellular respiration carried out by microorganisms. Microalgae and bacteria are supposed to reach a "natural" coexistence with one another, which takes into account the parameters of the reactor.

Nevertheless, the make-ups of the consortium that has reached the equilibrium may be very different from each other based on the variables that are currently present inside the reactor. The make-up of the consortium has a direct impact on the proportions between the numerous phenomena, such as the formation of oxygen, the consumption of carbon dioxide, and the assimilation of phosphorus and nitrogen. As a result, the levels at which these operations are carried out are constantly adapting to reflect the shifting composition of the consortium [16]. According to the researchers, the effectiveness of microorganisms used in the treatment of wastewater could be enhanced by eliminating the oils and fats that are contributing to the degradation of the habitat and by trying to manipulate the existence of relevant species of bacteria used in cold or hot water treatment.

This chapter provides an overview on the necessity of the treatment of wastewater before offering a look into the most recent developments in the autonomous utilisation of bacteria and microalgae for the treatment of wastewater, then concludes with an explanation of the implementation of microalgal species and bacteria that are heterotrophic substances, incorporated together with a method of treating wastewater that is less harmful to the environment based on scientific evidence observed in this arena of research.

## 6.2 IMPLICATIONS OF TREATING WASTEWATER IN PRESENT SCENARIO

Industrial effluents are harmful to human beings, other animals, and the environmental systems. Infrastructure is needed for the purification of waste-stream aids to purify the water and alleviate conditions like those that are prevalent in developing countries at this present era [17, 18].

Water that has not been treated creates major health concerns and is responsible for 1.7 million fatalities per year; more than 90% of these deaths occur in developing nations. In several developing nations, a smaller proportion of household and urban effluent is addressed before being released into the environment. Thus, a number of water-related diseases, such as cholera and schistosomiasis, continue to be widespread. According to the researchers, the existence of heavy metals in aqueous effluent can cause significant health problems, such as cardiovascular illnesses, damage to neural tissue, renal damages, as well as cancer and diabetes [19]. Even though Mother Nature tries very hard to process the waste stream in a natural way, there is simply far too much for it to manage on its own [20, 21]. Researchers have shown that the amount of wastewater being produced is steadily growing in parallel with the expansion of the planet's population.

The aerobic wastewater treatment systems of today take advantage of the ambient air flow that is present. Because the circumstances necessary for the full oxidation of the organic contaminants into $CO_2$, $N_2$, and $H_2O$ are generated by this procedure, the organic contaminants and the odours that are connected with them are removed. Since the effluent has been treated, it is no longer contaminated with pollutants and can be released back into the environment [22]. The biological process is being overwhelmed by the presence of billions of individuals. If wastewater was not properly treated, the volume of wastewater could inflict devastation, much as it does nowadays in regions that are yet in the developing stages of their economies. Over 80% of the world's wastewater is released into the environment without first being treated. Countries that have water treatment facilities employ a variety of approaches to purify wastewater before releasing it back into the natural surroundings [23–25]. This is done with the intention of ensuring the continued health and well-being of both humans and the planet [23–25].

## 6.3 MULTIPLE STREAMS OF WASTEWATER FROM THE INDUSTRIAL SECTOR

### 6.3.1 BATTERY INDUSTRY

Generally, battery industries are known as the most contaminated and dangerous industry that produces numerous amounts of impurities. The structure of batteries consists of a negative electrode, a positive electrode, and a solution containing electrolytes. Grid casting, milling of oxide, pasting of the grid, curing of the plate, plate parting of the plate, and finally the assembly of battery are steps involved in the production of battery [26]. Water is utilised throughout the production process, both for the preparation of reactive chemicals and electrolytes as well as for the deposition of reactive components on the surface. As a direct result of this, wastewater is produced, which can be identified by its high concentrations of Cd, Ni, and Ag. Additionally, the composition of this wastewater is determined by the specific procedure that is implemented in the production of batteries.

## 6.3.2 ELECTRICAL POWER PLANT

One of the primary manufacturing sectors is electric power plants engaged in supplying the power and electricity to urban areas. Generally, electric power plants have only a few major unit processes. Initially, the fuel resource is delivered to the power plants, where they need to be processed to improve productivity and then kept in a proper way. After the fuel burning process, the heat generated is transferred to the boilers that consequently alters the solution into steam [27]. Accordingly, the steam generated is circulated inside the turbine where the rotation of turbine initiates to produce electricity. Last, the steam is delivered to the condenser. The outlet stream condenses inside the steam turbine to form purified water which is periodically returned to the boiler. Condensation of steam is one of the phases that results in wastewater effluents. Although the wastewater discharge characteristics do not surpass the limit of the sewage system, they do outstrip the limit of irrigation, which is cause for concern. Figure 6.3 shows the methods adopted for treating effluent produced by electric power plant.

## 6.3.3 NUCLEAR POWER PLANT

The effluents from nuclear-based power plants contain radioactive materials that are emitted in liquid and airborne forms. Nuclear fission is the primary cause for the discharge of airborne effluents and the activation of tritium ($H_2$) gases. These effluents are discharged by both boiler water reactors (BWR) and pressurised water reactors (PWR) [28]. The management of radioactive waste, effluent control systems, and analytical techniques used to monitor the effluent are designed and operated in a way that causes instability in airborne effluent. Before being permitted for release to the sea, radioactive liquid wastes discharged into the aqueous streams are under observation. Additionally, groundwater pollution has been caused by unchecked leakages of molten radioactive wastes brought on by fission activity. For PWR compared to BWR, the activity of tritium in liquefied wastes is substantially greater. Due to the mixed activation and fission products, nuclear facilities currently regularly emit a limited curies of the tritium ($H_2$) in molten wastes.

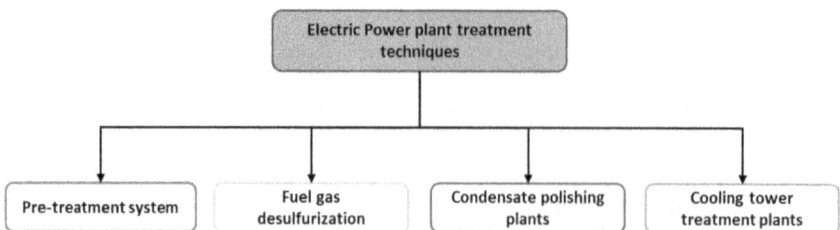

FIGURE 6.3  Electric Power Plant Wastewater Treatment Technologies.

### 6.3.4  TEXTILE INDUSTRY

Historically, textile industry has a long present in human life. Indeed, it represents the principal and firstborn industry worldwide. In general, the textile industry involves the use of highly toxic chemicals that represent 70% of the discharged wastewater. The wastewater management methods used in textile industries include the elimination of suspended solids with improvement of the quality of effluent using sedimentation, screening, neutralisation, chemical coagulation, and mechanical flocculation [29].

### 6.3.5  PULP AND PAPER INDUSTRY

Generally, pulp and paper industries generate enormous amounts of biomass residues [30]. The methods used for the paper and pulp industry include debarking, chipping, pulping, and bleaching. Initial treatment methods such as sedimentation, flotation, and filtration are used to treat paper and pulp wastewater. Secondary management (biological) includes aerated lagoons, activated sludge, and anaerobic treatment (Acidogenesis, Hydrolysis, Methanogenesis and Acetogenesis). Last, tertiary treatment includes ozonation, precipitation, and membrane separation (Microfiltration -MF, Ultrafiltration-UF, Nanofiltration-NF, and Reverse Osmosis-RO).

### 6.3.6  LEATHER INDUSTRY

The leather industry is consistently recognised as the most profitable industry worldwide. However, it is also considered one of the most polluting in the world. The production of leather has a staggeringly large number of adverse effects on the surrounding ecosystem as a result of its existence. The polished leather is used for a wide variety of products, some examples of which are footwear, clothes, and suitcases. In order to accomplish this, large quantities of raw materials are collected, and then they are subjected to various chemical processes in order to make the final products [31].

### 6.3.7  AGRICULTURAL INDUSTRY

In general, the modern agriculture business requires a significant amount of freshwater from environmental resources, and it also generates an enormous volume of effluent. In most cases, the pollutants that make up wastewater include things like organic and inorganic materials (such as soluble minerals), nutrients (such as phosphorous, nitrogen, and potassium), toxicants, and germs [32]. Nutrients, manure, suspended particles, viruses and bacteria, biochemical oxygen demand (BOD), chemical oxygen demand (COD), and pesticides are removed using a variety of filtration procedures throughout the methods of treatment that are utilised in the agricultural industry. Other methods include mechanical techniques such as sedimentation, filtration, separation, crystallisation and flotation, and biological methods including activated sludge treatment under anaerobic environments and physico-chemical techniques such as coagulation, electro-coagulation, and ozonation and flocculation [33].

### 6.3.8 Pharmaceutical Industry

It is generally agreed that the pharmaceutical industries are an essential sector for every nation. Despite this, the industry is responsible for a significant portion of the pollution that has been caused because it uses around 99% of water in its manufacturing. The medication production sector is the primary contributor to the untreated wastewater from this sector. This industry includes a variety of hazardous components that are harmful to humans as well as animals. Several different treatment methods, such as reverse osmosis (RO), membrane filtration, coagulation, and flocculation, are utilised in order to bring about a decrease in the concentration of various constituents found in the effluent, including BOD, COD, and total dissolved solids (TDS) [34].

### 6.3.9 Oil Mill Industry

Oil used for cooking accounts for a significant portion of the world's total food consumption. Cleaning the seeds, pressing, grinding, extracting, deodourising, bleaching, pre-chilling, and packaging are all necessary steps in the production of cooking oil. These processes constitute the bulk of what mills do. Overall, the bleaching procedure and the ETP unit are the sources of the effluent [35]. Following the first treatment, the harmful by-products proceed through a series of further biological treatments.

### 6.3.10 Organic Chemical Production Industries

The organic chemical industry is the sector that has the largest negative impact on the surrounding environment. In most cases, this industry discharges wastewater that contains an exceptionally high concentration of potentially hazardous organic and inorganic pollutants [36]. The bulk of the pollutants that are released have been demonstrated to have features that make them pathogenic, cancerous, but not degradable. The most common wastewater treatment methods, including dispersed air flotation, gravity separation, de-emulsification, coagulation, skimming, and flocculation, are utilised in the process of chemical runoff purification.

### 6.3.11 Petroleum and Petro-Chemical Industries

The production of petroleum results in the release of both gaseous and liquid types of pollutants. It is far easier to exercise control over gaseous emissions than over the liquid contaminants. In order to minimise environmental contamination, Effluent Plant (ETP) will effectively remove liquid contaminants [37].

### 6.3.12 Mines Industry

In its broadest sense, mining describes the activity of taking mud, sands, rocks, and additional materials from the soil. The location of the mining is referred to as a quarry. Most of the time, this industry pollutes the area's natural resources [38]. Most of the time, very toxic wastewater is released by quarries and mines. These extremely hazardous metals affect the water resources and cause several ailments. Chemical oxidation, neutralisation, biological pre-treatment, and co-precipitation are some of the processes used to treat the effluents that are produced.

### 6.3.13 Dairy Industry

The dairy sector generates a lot of wastewater, and that wastewater comprises a significant quantity of particulate organic substances. The vast majority of milk that is processed by the dairy sector contains significant quantities of various milk components, including inorganic salts and casein [39]. The efficacy of the principal physiochemical processes utilised in the dairy sector for treatment processes has been measured to be as high as 98%., Coagulation/flocculation, precipitation, membrane treatment, and other methods are all included in these procedures. The properties of dairy wastewater are presented in Table 6.1.

### 6.3.14 Steel and Iron Industry

It is believed that the effluent produced by the facilities that process coke oven by-products is the most hazardous wastewater in the iron and steel sector because it encompasses poisonous compounds like cyanide phenol and ammonia. Untreated waste effluents from the steel production are believed to have various detrimental impacts, involving poisoning the aquatic species, a lessening in dissolved oxygen (DO), and water logging owing to suspended particulates. Phenol and $NH_3$ produced in the waste effluent raise the pH of the aqueous stream, which causes the toxicity. As a consequence of the introduction of degrading organic constituents into the aqueous stream, the organic matter is used as a source of carbon by the soil bacteria, which in turn contributes to a reduction in the DO concentration in effluents. The dangers posed by the effluents produced by this industry are outlined by Table 6.2.

### 6.3.15 Food Industry

Because it contains a significant level of both chemical oxygen demand (COD) and biological oxygen demand (BOD), effluent from the food industry is a major contributor to environmental pollution. Because water is required in the majority of plant

## TABLE 6.1
### Properties of Dairy Sector Effluents

| Parameters | Effluent Released | Average | Ref. |
|---|---|---|---|
| BOD (mg/L) | 1300 | 100 | [40] |
| COD (mg/L) | 2400 | 250 | [41] |
| pH | 1–5.8 | 6.5–8.0 | [42] |
| Temp (C) | 25–50 | 33–38 | [43] |
| TDS (mg/L) | 5600 | 2100 | [44] |
| Oil and Grease (mg/L) | 35 | 10 | [41] |
| TSS (mg/L) | 700 | 100 | [45] |

**TABLE 6.2**
**Toxicity Impact of Steel and Iron Industries**

| Type of Contaminant | Effect | Disinfection | Ref. |
|---|---|---|---|
| Aromatic Hydrocarbons | Multiple impacts depending on the type of PAH | Mutagenesis and carcinogenesis can be caused by non-polar narcosis and photo-toxicity. | [46] |
| Cyanides | Have effects that change over time based on the dosage and the speed. | Toxic effects that can be fatal if hydrogen cyanide gas is breathed in. | [47] |
| Heavy metals | Influence the body's enzyme system as well as its metabolic rate. | Damages in the reproduction system, the circulatory, as well as the cardiovascular systems respectively. | [45] |
| Fluorides | Even at low quantities, it is possible to stop the development enzyme phosphatase from working. | Diseases such as osteoarthritis and gout, in addition to cancers and brain injury. | [48] |
| Surfactants | Enhance the bioavailability. | A detrimental effect on the ability of heterotrophic nano-flagellate to survive. | [46] |

activities, the food industry consumes a far higher quantity of it than does in any other sectors. According to the findings, the levels of BOD/COD, total solids, and colloidal solids in the effluents of these companies fluctuated over time [49]. This is due to the fact that various culinary products call for a variety of distinct ingredients. The chocolate manufacturing process is among the most water-intensive and therefore among the most polluting processes in the food industry.

## 6.4   THE HYBRID MECHANISM OF MICROALGAL/BACTERIAL APPROACH TOWARDS THE PURIFICATION OF WASTEWATER

The hybrid bacterial–microalgal strategy is becoming a focal point in industrial zones as a means of enhancing the environmental status of water reserves and, more specifically, reducing the amount of phosphorus (P) and nitrogen ($N_2$) that is present in industrial wastewaters. This interest has been spurred by the need to improve the ecological condition of water resources.

The capability of microalgae to utilise the biological and chemical carbon, together with the synthetic phosphorus (P) and nitrogen (N), in wastewater for their population increase, with the intended result of a decrease in the accumulation of these intoxicants in the water, necessitates the implementation of microalgal species [50].

It was approximated that the plant matter of phytoplankton biomass manufactured from the release of 1 kg of P could indeed elicit 100 kg of $O_2$ supply, whereas that produced from the outflow of 1 kg of N can assert 14 kg of $O_2$ supply. As a result of this lack of oxygen or anaerobic environments, species composition is ended up losing, and the overarching capability of the aquatic environment is significantly affected [51].

It was determined the eutrophication-inducing threshold level to be in the band of 0.21–1.2 mg L1 for total nitrogen (TN) and 0.01–0.1 mg L1 for total phosphorus (TP), correspondingly. This range of concentrations was discovered to be responsible for producing eutrophication [52]. In most cases, phosphorus is thought to be the nutrient that has the most impact on the rate of phytoplankton development. As a result, reducing eutrophication requires a reduction in the amount of phosphorus that is introduced into receiving networks [53]. The difficult process of removing nitrogen and phosphorus from the water demands a distinct environment, which in turn requires a significant amount of energy. As a direct consequence of this, the total expense of the processing goes up significantly [51]. Figure 6.4 illustrates the hybrid mechanism of microbial community for wastewater treatment.

According to the researchers, the use of eukaryotic algal species and cyanobacteria is recognised as a viable alternative to energy-intensive and traditional biochemical treatment techniques that are also environmentally benign [54, 55]. It was found that the utilisation of microalgae in the wastewater treatment procedure is a method for the bio-fixation of $CO_2$ that is not only possible but also cost effective. In addition, it is a renewable source for biomass [56].

It has been established that the strain *Scenedesmus obliquus* was successful in removing nutrients (N, carbon, and P) from piggery effluent [57]. *Chlorella pyrenoidosa* was shown to be effectively grown in dairy waste effluent following acclimation with wastewater [58]. This was accomplished while simultaneously lowering the concentration of mineral and chemical components such as TS (total solids) and Alkalinity. There have been reports of certain species of Chlorella, such as *C. vulgaris*, exhausting P and N from the grey water effluent during the initial level of the diagnosis. It was demonstrated that the use of *C. vulgaris* was capable of successfully eliminate the indicated 88% biological oxygen demand (BOD), 82% total nitrogen, and 54% total phosphorus from the original amounts that were present in the brewing effluent [59]. *Chlamydomonas sp.*, *Dunaliella sp.*, *Spirulina sp.*, and *Botryococcu sp.* are some of the other species of micro-algae that have been investigated for their phytoremediation potential [60]. Other micro-algae species include the amount of nutrients utilised by the microalgae directly connected with its development [61]; reducing the availability of these nutrients can likewise limit or lower the growth of the microalgae. Figure 6.5 exhibits an up-flow anaerobic sludge blanket reactor for wastewater treatment.

## 6.5 THE ADVANTAGES AND DISADVANTAGES OF THE HYBRID APPROACH USING MICROALGAL/BACTERIAL TREATMENT FOR WASTEWATER PURIFICATION

The holistic strategy's practicality provides a few advantages [62]. Among these are the following: (a) the capability of removing nutrition and organic carbonaceous substances in a single component; (b) the systems have a lesser stipulation for physical aeration, which results in a decreased carbon emissions; (c) the generation of

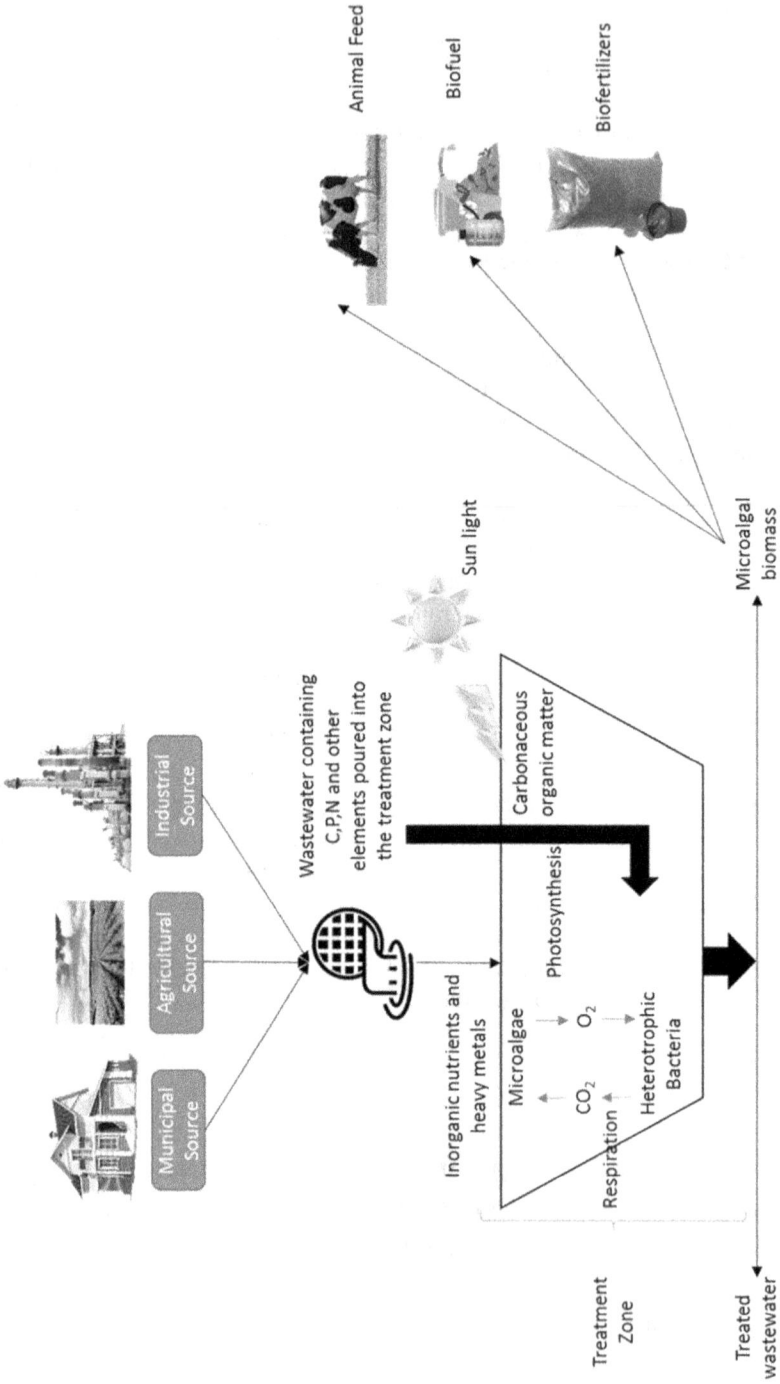

**FIGURE 6.4** Hybrid Mechanism of Microbial Community for Wastewater Treatment.

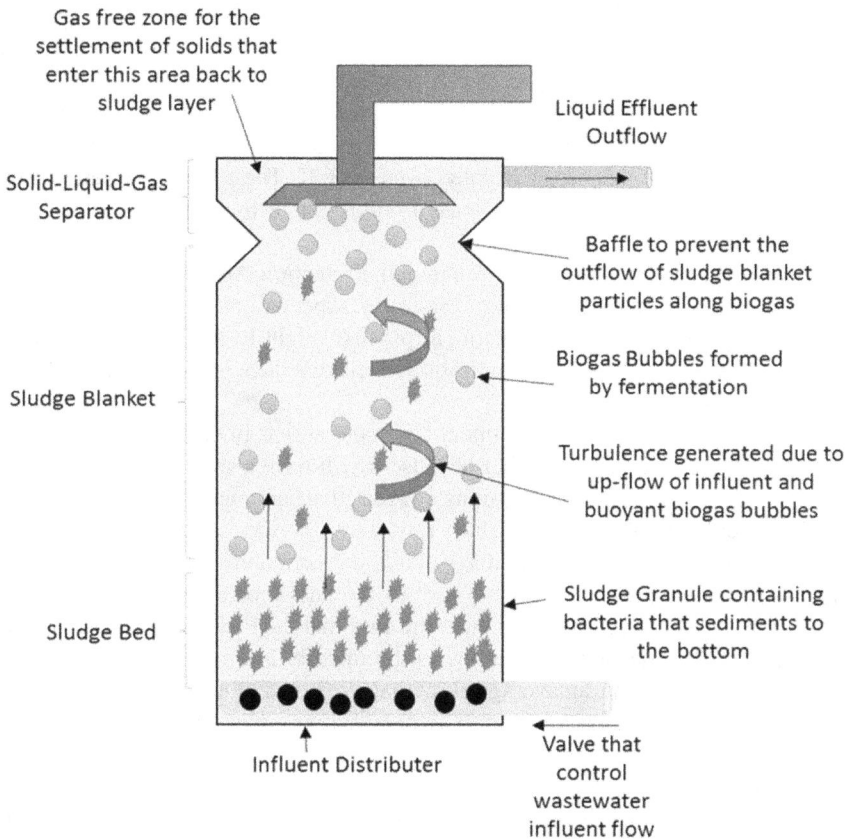

Gas free zone for the settlement of solids that enter this area back to sludge layer

Liquid Effluent Outflow

Solid-Liquid-Gas Separator

Baffle to prevent the outflow of sludge blanket particles along biogas

Biogas Bubbles formed by fermentation

Sludge Blanket

Turbulence generated due to up-flow of influent and buoyant biogas bubbles

Sludge Granule containing bacteria that sediments to the bottom

Sludge Bed

Influent Distributer

Valve that control wastewater influent flow

**FIGURE 6.5** Up-Flow Anaerobic Sludge Blanket Reactor Using Bacterial Species for Wastewater Treatment.

resilient biomass in the manifestation of bio-residues that could be utilised for the biodiesel production or synthesis of biogas; (d) the generation of smaller quantity of sludge in contrast to standard treatment innovations that consume bacteria.

In addition, the researchers have discovered that this method has a few drawbacks as well. These drawbacks include: (a) the light reliance of microalgae cultivation that is not accurate for genuine activated sludge based procedure; (b) the higher pH of the medium produced because of photosynthesis by microalgal species that has a detrimental effect on the group of bacteria affiliated with activated sludge based process; and (c) the slower growth percentage of microalgal species that gives rise to greater solid retention span/time (SRT).

In addition to these obstacles, the wastewater treatment using microalgae is faced with other issues, one of which is the management of the vast amounts of effluent which need to be processed. Thus, the various microalgae growing procedures need to be examined in order to come up with (a) optimum microalgal production, (b) assuring the greater efficiency of the nutrient extraction, and (c) how to deal with extremely high numbers of different microalgal species. The construction and layout of the reactors, which control factors such as temperature and light and have an influence on the development of microalgae, also has a significant impact on the wastewater treatment process. It was found that the various strategies for cultivating microalgae can be roughly divided into two groups' suspended systems and immobilised systems. One of the two systems is then designated as an open platform while the other is an isolated system [63]. The synergistic interactions for utilising bacteria and algal biomass is represented by a simplified layout of Figure 6.6.

According to the Swedish Environmental Protection Agency (2008), in order to prepare for the implementation of the lower expectations that have been established, the achievement of the treatment process ought to be capable of meeting the current requisite effluent concentration levels that have been established by the urban water treatment guidelines.

It is grander for dropping investment costs and assists in optimising the contact area needed if the device is smaller because it will have a shorter hydraulic retention time [64]. Numerous conclusions suggest that, among the immobilised and suspended cultivation process, a short processing period is noticed in photo-bio reactor based processes, with greater P and N extraction efficacy regardless of the wide range of operational variables, such as biomass inoculum density, and heat, in addition to solar irradiation. This was the case that the PBR suspended processes were subjected to the same conditions as the immobilised cultivation systems. In PBR suspended systems, an average percentage removal of 87.3% for nitrogen and 82.9% for phosphorus was attained in an overall time span of 3.1 days [59].

The combination of bacterial and algal substrate for the wastewater treatment process has been shown to be a viable solution both from a technical and an environmental point of view in recent studies. When compared to the traditional method of treating wastewater, the employment of microalgae as an alternative for the elimination of nitrogenous, phosphorous, and organic carbon material leaves a smaller footprint in regard to the amount of energy consumed and the amount of greenhouse gases produced. Because a centralised wastewater treatment plant has never been capable of keeping up owing to the demand of providing accommodations to a growing populace in urban regions, decentralisation of treating wastewater is becoming essential in modern context. The researchers are additionally reluctant to increase their existing capacity because the majority of their locations are in cities where space is at a premium and so cannot accommodate further buildings. Earlier a research group conducted research on the effectiveness of utilising an integrated strategy when dealing with this issue [62]. The assignment of algal biofilm communities to a previously established activated sludge procedure (ASP) premised decentralised waste-effluents treatment plant (DWWTP) in Indian regions which results in a substantial augmentation of carbonaceous substrate and nutrient expulsion from the effluent stream within a solitary reactor. In order to achieve this goal, the algae were immobilised using a method that was appropriate for the situation, and they were given the ideal amount of sunlight and solar radiation [62].

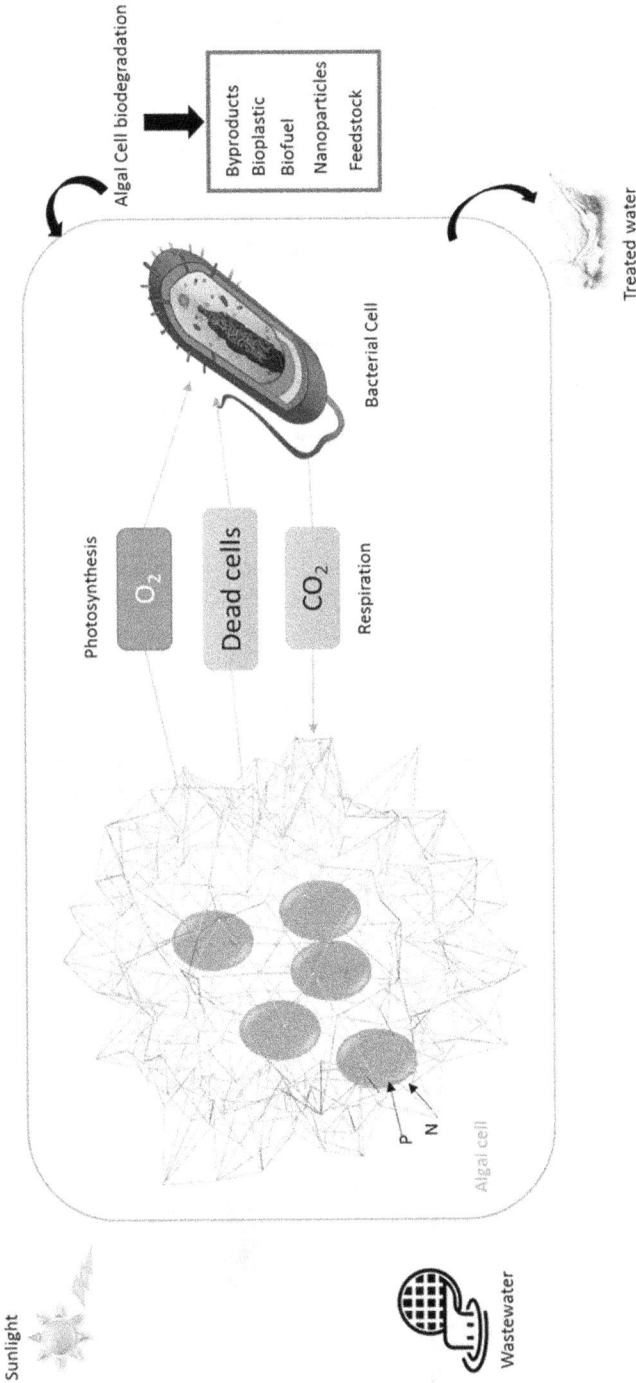

**FIGURE 6.6** The Simplified Layout of Synergistic Interaction for Hybrid Bacteria/Algal System for Wastewater Treatment.

## 6.6  CONCLUSION

The percentage of the earth's surface that is covered by water is approximately 70%. Of this 70%, approximately 97.5% of the water is found in oceans and seas and the remaining 1.73% is located in the form of ice. Only 0.77 % of the fresh water on the surface of the world is usable for commercial and farming purposes [65] (Figure 6.7). According to the World Health Organization, around 1.7 million people have lost their lives as a direct consequence of water pollution, and approximately 4 billion symptoms of varied health disorders are recorded annually as a direct outcome of water-borne diseases [66]. The statistics clearly reflects the importance of wastewater treatment.

Of the wastewater that is generated in developing nations, 50–80 % is either not processed at all or is just inadequately treated. As a direct result of urbanisation and the continuous rise in the world's population, the volume of wastewater is continuously growing. The standard sewage treatment technologies, such as activated sludge, are not effective in reducing the nutrient levels of effluents to levels that are suitable for preventing an accelerated pace of algal blooms in the area surrounding sewage dumping sites. The incorporation of microalgae into the sewage treatment process has a wide range of potential benefits, including (a) the expulsion of excessive amounts of nitrogen source, phosphorus, and heavy metals; (b) the prevention of the proliferation of bacterial pathogens; (c) the microalgal-mediated sequestration of carbon dioxide; and (d) the generation of biofuels from microalgal biomass. The unified bacterial and microalgal strategy is proving to be further encouraging even though the heterotrophic bacteria have the ability to breakdown organic compounds in the utter lack of oxygenated atmosphere due to the fact that $O_2$ is supplied by the photosynthetic activity of the microalgae. In addition, the requirement for physical mixing of $CO_2$ is also completely eradicated due to the fact that $CO_2$ is generated by the respiration of the bacteria. Because of this, the integrated method of treating

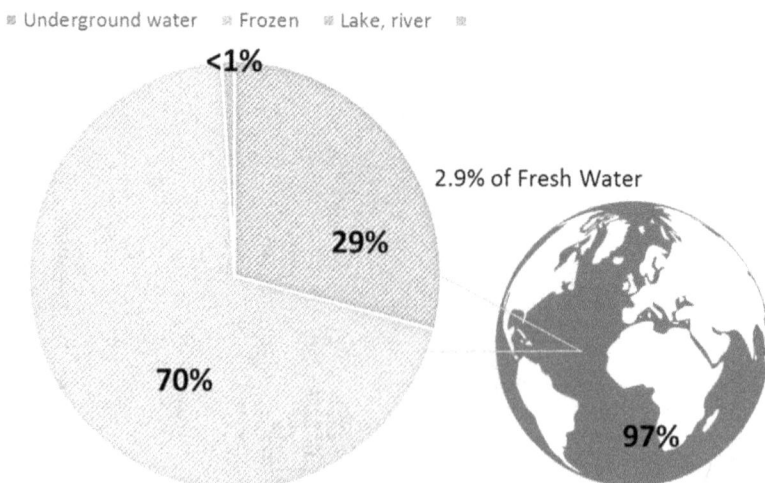

**FIGURE 6.7**  Distribution of Earth's Water.

wastewater using bacteria and microalgae would demand processing in a single step, which would reduce the amount of complexity involved in the remediation process.

Since the initial move towardsvincorporating microalgae in wastewater treatment plants, a significant number of scientific endeavours have been dedicated for developing this potentially useful application. Microalgae, which are employed in wastewater treatment plants, are a cost-effective method that yield various advantages and can play an outstanding role in these facilities. Microalgae have a wide range of techniques at their disposal, which allows them to eliminate a wide variety of contaminant classes and toxic waste, including those that are produced by domestic andvagricultural drainages and effluents, as well as the printing, pharmacological, and metal plating industries. The microalgal bioremediation process does not result in the production of secondary pollutants. In addition, the microalgal biomass that is discarded can be utilised as a source of source materials for the production of medications, biodiesel, fertilisers, and nourishing foods, all of which have considerable financial benefits. In spite of the fact that previous research has demonstrated that microalgae-based wastewater treatment plants are superior to traditional wastewater treatment plant technologies in terms of the elimination of nutrients, the capture of carbon, and the generation of biomaterials, their wide spectrum is still hindered by a number of obstacles. Because of this, further research is required for application of microalgae in the bio-refinery processes.

Microalgae use the $CO_2$ produced by heterotrophic bacteria to yield a greater amount of oxygen through photosynthesis. Carbohydrates are required for the generation of biomass, and oxygen is the terminal acceptor for aerobic cellular respiration carried out by heterotrophic bacteria. This cycle continues until the $CO_2$ produced by heterotrophic bacteria is used up by microalgae. Therefore, autotrophic protists and heterotrophic bacteria have the possibility to supplement each other, thereby making them simple to manipulate and risk-free for human consumption and application. This recurring, an almost self-sufficient approach to intervention with the additional advantage of microalgal biomass might deliver the appropriate combination between the expenses and the availability of hygienic water for the inhabitants in an area where wastewater management is uncommon, as well asenvironmental fortification criterions that are not rigorous and where water-borne pathogens are very commonly found. Unlike traditional water treatment, which requires a large-scale centralised plant, a technology that satisfies the requirements of being environmentally friendly, user-friendly, and cost-effective can be implemented in a decentralised way at the point of origin. In light of the growing quantity of wastewater load and with the intention of maximising the virtual quantity of microalgae and bacterial consortia, additional investigation needs to be conducted in this field. This is necessary in order to achieve good biomass yields. It requires a significant knowledge of the interactions among the essential species of bacteria and microalgae to stabilise treatment processes, as well as the growth encouraging elements for each of these, such as pH level, temperature, and nutrient concentration. For optimal performance, it is necessary to establish methods for the monitoring and provision of a constant intake of nutrients and sunlight (mostly within closed systems). High concentrations of meiobenthos have been found in areas where wastewater is discharged, and it is well known that these organisms feed on organic material that has dropped to the bottom.

In addition to bacteria and microalgae, the idea of including meiobenthic fauna – defined as benthic metazoans which are able to flow through a filter with a size of 500 mm and yet are trapped by one with a size of 40 mm – could be also investigated. This can be explained by the fact that they are able to consume particle organic matter for food, which in turn accelerates the cycle of organic matter decomposition.

They can also serve as preadators for the bacteria that form biofilms inside the filter bed that are employed for the mechanical wastewater treatment, which would provide an expected alternative method for cleaning the filters. Instead of trusting on chemical and physical decontamination processes such as chlorine or UV treatment, for example, the bacterial biomass that is produced during the treatment method could also be directed towards these meiobenthos. This would result in a natural reduction in the number of meiobenthos that are present in the effluent. Nevertheless, the implementation of such disinfection procedures might lead to better results.

## ACKNOWLEDGMENTS

The authors are thankful for the funding provided by Malaysian Joint Research Scheme-ST 077–2022, Interdisciplinary Research IIRG003A-2022IISS and International Grant ICF 023–2022 and ICF 080–2021 under University of Malaya, Kuala Lumpur 50603, Malaysia.

## REFERENCES

[1]  Mastropetros, S. G., Pispas, K., Zagklis, D., Ali, S. S., and Kornaros, M. 2022. Biopolymers production from microalgae and cyanobacteria cultivated in wastewater: Recent advances. *Biotechnology Advances*. 60:107999.
[2]  Al-Tohamy, R., Ali, S. S., Li, F., Okasha, K. M., Mahmoud, Y. A., Elsamahy, T., Jiao, H., Fu, Y., and Sun, J. 2022. A critical review on the treatment of dye-containing wastewater: Ecotoxicological and health concerns of textile dyes and possible remediation approaches for environmental safety. *Ecotoxicology Environmental Safety*. 231:113160. doi: 10.1016/j.ecoenv.2021.113160.
[3]  Jouhara, H., Bertrand, D., Axcell, B., Montorsi, L., Venturelli, M., Almahmoud, S., Milani, M., Ahmad, L., and Chauhan, A. 2021. Investigation on a full-scale heat pipe heat exchanger in the ceramics industry for waste heat recovery. *Energy*. doi: 10.1016/j.energy.2021.120037.
[4]  Jun, K. C., Raman, A. A. A., and Buthiyappan, A. 2020. Treatment of oil refinery effluent using bio-adsorbent developed from activated palm kernel shell and zeolite. *RSC Advances*. 10(40):24079–24094. http://xlink.rsc.org/?DOI=d0ra0 3307c.
[5]  Upadhyay, K., and Srivastava, J. K. 2017. Application of ozone in the treatment of industrial and municipal wastewater. *Journal of industrial Pollution Control*. 21(1):235.
[6]  Upadhyaya, S., and Kynčlová, P. 2017. Big data-its relevance and impact on industrial statistics. *Scope Use Big Data Industrial Statistics*. 1:20–29. doi: 10.13140/ RG.2.2.31044.94083.
[7]  Zhuang, L. L., Yang, T., Zhang, J., and Li, X. 2019. The configuration, purification effect and mechanism of intensified constructed wetland for wastewater treatment from the aspect of nitrogen removal: A review. *Bioresource Technology*. 2019;293:122086. doi: 10.1016/j.biortech.2019.122086.
[8]  Tiffon, C. 2018. The impact of nutrition and environmental epigenetics on human health and disease. *International Journal of Molecular Science*. 19(11):3425.

[9] Gaur, V. K., Sharma, P., Sirohi, R., Awasthi, M. K., Dussap, C. G., and Pandey, A. 2020. Assessing the impact of industrial waste on environment and mitigation strategies: A comprehensive review. *Journal of Hazardous Materials*. 398:123019.

[10] Etim, M. A., Babaremu, K., Lazarus, J., and Omole, D. 2021. Health risk and environmental assessment of cement production in Nigeria. *Atmosphere*. 12(9):1111.

[11] Alexandre, C. L., Antonio, V., Diego, L., and Cristina, G. F. 2018. Energy balance and life cycle assessment of a microalgae-based wastewater treatment plant: A focus on alternative biogas uses. *Bioresource Technology*. 270:138–146. doi: 10.1016/S0960–8524(00)00136.

[12] Gudiukaite, R., Nadda, A. K., Gricajeva, A., Shanmugam, S., Nguyen, D. D., and Lam, S. S. (2021). Bioprocesses for the recovery of bioenergy and value added products from wastewater: A review. *Journal of Environmental Management*. 300:113831. doi: 10.1016/j.jenvman.2021.113831.

[13] Arora, K., Kaur, P., Kumar, P., Singh, A., Patel, S. K. S., and Li, X., et al. (2021). Vaporization of wastewater resources into biofuel and value-added products using microalgal system. *Frontiers Energy Research*. 9:646571. doi: 10.3389/fenrg.2021.646571.

[14] Dhanker, R., Hussain, T., Tyagi, P., Singh, K. J., and Kamble, S. S. 2021. The emerging trend of bio-engineering approaches for microbial nanomaterial synthesis and its applications. *Frontiers Microbiology*. 12:638003. doi: 10.3389/fmicb.2021.638003.

[15] Muñoz, R., and Guieysse, B. 2006. Algal-bacterial processes for the treatment of hazardous contaminants: A review. *Water Research*. 40:2799–2815. doi: 10.1016/j.watres.2006.06.011.

[16] Molazadeh, M., Ahmadzadeh, H., Pourianfar, H. R., Lyon, S., and Rampelotto, P. H. 2019. The use of microalgae for coupling wastewater treatment with $CO_2$ biofixation. *Frontiers Bioengineering Biotechnology*. 7:42. doi: 10.3389/fbioe.2019.00042.

[17] Desai, C., Jain, K. R., Boopathy, R., van Hullebusch, E. D., and Madamwar, D. 2021. Editorial: Eco-sustainable bioremediation of textile dye wastewaters: Innovative microbial treatment technologies and mechanistic insights of textile dye biodegradation. *Frontiers Microbiology*. 12:707083. doi: 10.3389/fmicb.2021.707083.

[18] Benedetti, M., Vecchi, V., Barera, S., and Osto, L. 2018. Biomass from microalgae: The potential of domestication towards sustainable biofactories. *Microbial Cell Factories*. 17:173. doi: 10.1186/s12934-018-1019-3.

[19] Jaishankar, M., Tseten, T., Anbalagan, N., Mathew, B. B., and Beeregowda, K. N. 2014. Toxicity, mechanism and health effects of some heavy metals. *Interdisciplinary Toxicology*. 7:60–72. doi: 10.2478/intox-2014-0009.

[20] Chia, Y. C., Tzu-Yao, W., and Gen-Schuh, W. 2007. Determining estrogenic steroids in Taipei waters and removal in drinking water treatment using high-flow solid-phase extraction and liquid chromatography/tandem mass spectrometry. *Science Total Environment*. 378:352–365. doi: 10.1016/j.scitotenv.2007.02.038.

[21] Cristine, S., Benjamin, F., Kirsten, L., Luis, E. M., Georg, S., and Martin, H. 2011. Sodium overload and water influx activate the NALP3 inflammation. *Biological Chemistry*. 286:35–41.

[22] Marina, K., Sarah, T., Khalef, R., Shareef, J., Gaby, K., and Racheed, A. 2019. Sources of microplastics pollution in the marine environment: Importance of wastewater treatment plant and coastal landfill. *Marine Pollution Bulletin*. 146:608–618.

[23] Hoh, D., Watson, S., and Kan, E. (2016). Algal biofilm reactors for integrated wastewater treatment and biofuel production: A review. *Chemical Engineering Journal*. 287:466–473. doi: 10.1016/j.cej.2015.11.062

[24] Holger, D., Michael, W. T., and Michael, W. (2006). Wastewater treatment: A model system for microbial ecology. *Trends Biotechnology*. 24:483–489. doi: 10.2174/18722 08313666190924162831.

[25] Hussain, J., Wang, X., Sousa, L., Ali, R., Rittmann, B. E., and Liao, W. (2020). Using nonmetric multi-dimensional scaling analysis and multi-objective optimization to evaluate

green algae for production of proteins, carbohydrates, lipids, and simultaneously fix carbon dioxide. *Biomass Bioenergy*. 141:105711. doi: 10.1016/j.biombioe.2020.105711.

[26]  Rahangdale, R. V., Kore, S. V., and Kore, V. S. 2012. Waste management in lead acid battery industry: A case study. *World Journal of Applied Environmental Chemistry*. 1:7–12.

[27]  Srikanth. 2015. Wastewater effluent analysis for electric power plants. *Water Science Technology*. 62(10):2256–2262. doi: 10.1186/s40643-017-0163-7.

[28]  Efremenkov, V. M. 2014. Liquid effluents released in nuclear industry. *International Journal of Renewable Energy Technology Research*. 3(7):1–10. doi: 10.1007/s10582–003–0075-y.

[29]  Khandegar, V., and Saroha, A. K. 2013. Electrocoagulation for the treatment of textile industry effluent–a review. *Journal of Environmental Management*. 128:949–963. doi: 10.1016/j.jenvman.2013.06.043.

[30]  Hubbe, M. A., Metts, D., and Blanco, Z. K. 2016. Wastewater treatment and reclamation: A review. *BioResource*. 11(3):7953–8091. doi: 10.2175/106143010X12851009156321.

[31]  Biswas, T., and Chowdhury, M. 2013. Treatment of leather industrial effluents by filteration and coagulation processes. *Water Resource Industry*. 3(1):11–22. doi: 10.1016/j.wri.2013.05.002.

[32]  Levy, G. J., Fine, P., and Bar-Tal, A. 2011. *Treated wastewater in agriculture*. Blackwell Pub. doi: 10.1002/9781444328561.

[33]  Oron, G., DeMalach, Y., Hoffman, Z., and Manor, Y. 2012. Effect of effluent quality and application method on agricultural productivity and environmental control. *Water Science Technology*. 26:7–1601. doi: 10.2166/wst.1992.0603.

[34]  Azizi, E., Fazlzadeh, M., Ghayebzadeh, M., Hemati, L., Beikmohammadi, M., Ghaffari, H. R., Zakeri, H. R., and Sharafi, K. 2017. Application of advanced oxidation process ($H_2O_2$/UV) for removal of organic materials from pharmaceutical industry effluent. *Environmental Protection Engineering*. doi: 10.5277/epe170115.

[35]  Rajkumar, K., Muthukumar, M., and Sivakumar, R. 2010. Novel approach for the treatment and recycle of wastewater from soya edible oil refinery industry—an economic perspective. *Resource Conservation Recycling Advances*. 54(10):52–758. doi: 10.26872/jmes.2018.9.1.3.

[36]  Ferraz, F. M., Povinelli, J., and Vieira, E. M. 2013. Ammonia removal from landfill leachate by air stripping and absorption. *Environmental Technology*. 34(15):2317–2326.

[37]  Saha, P. D., Baskaran, D., Malakar, S., and Rajamanickam, R. 2015. Comparative study of biofiltration process for treatment of VOCs emission from petroleum refinery wastewater-a review. *Environmental Technology Innovation*. 8:441–461. doi: 10.1016/j.eti.2017.09.007.

[38]  Kuyucak, N. 2008. Mining, the environment and the treatment of mine effluents. *International Journal of Environmental Pollution*. 10(2):315–325. doi: 10.1504/IJEP.1998.005151

[39]  Raghunath, B. V., Punnagaiarasi, A., Rajarajan, G., Irshad, A., and Elango, A. 2016. Impact of dairy effluent on environment—a review. *Integrated Waste Management System India*. doi: 10.1007/978-3-319-27228-3_22.

[40]  Kolhe, A. S., Ingale, S. R., and Bhole, R. V. 2009. Effluent of dairy technology. Shodh Samiksha Aur Mulyankan. *International Research Journal*. 2:459–461.

[41]  Ahmad, T., Aadil, R. M., Ahmed, H., Rahman, U., Soares, B. C., and Souza, S. L. 2019. Treatment and utilization of dairy industrial waste: A review. *Trends of Food Science and Technology*. 88:361–372. doi: 10.1016/j.tifs.2019.04.003.

[42]  Shete, B. S., and Shinkar, N. P. 2013. Dairy industry wastewater sources, characteristics & its effects on environment. *International Journal Current Engineering Technology*. 3(5):611–1615. doi: 10.1177/01445 98717698081.

[43] Zhang, H., Feng, J., Chen, S., Zhao, Z., Li, B., Wang, Y., Jia, J., Li, S., Wang, Y., Yan, M., and Lu, K. 2019. Geographical patterns of nirS gene abundance and nirS-type denitrifying bacterial community associated with activated sludge from different wastewater treatment plants. *Microbial Ecology.* 77(2):304–316. doi: 10.1007/s00248-018-1236-7.

[44] Tikariha, A., and Sahu, O. 2014. Study of characteristics and treatments of dairy industry waste water. *Journal of Applied Environmental Microbiology.* 2(1):16–22. doi: 10.12691/jaem-2-1-4.

[45] Sinha, S., Srivastava, A., Mehrotra, T., and Singh, R. 2019. A review on the dairy industry waste water characteristics, its impact on environment and treatment possibilities. *Emerging Issues Ecological Environmental Science.* doi: 10.1007/978-3-319-99398-0_6.

[46] Beh, C. L., Chuah, A. L., Nourouzi, M., and Choong, T. S. Y. 2014. Removal of heavy metals from steel making waste water. *E-Journal of Chemistry.* 9(4):2557–2564. doi: 10.1155/2012/128275.

[47] Bora, T., and Dutta, J. 2014. Applications of nanotechnology in wastewater treatment— a review. *Journal of Nanoscience Nanotechnology.* 14(1):613–626. doi: 10.14256/JCE.2165.2017.

[48] Das, P., Mondal, G. C., Singh, S., Singh, A. K., Prasad, B., and Singh, K. K. 2018. Effluent treatment technologies in the iron and steel industry-a state of the art review. *Water Environmental Research.* 90(5):395–408. doi: 10.2175/106143017X15131012152951.

[49] Noukeu, N. A., and Priso, R. J. 2016. Effluent treatment in food industry. *Water Resource Indus Basin.* 16(1):1–18. doi: 10.1016/j.wri.2016.07.001.

[50] Wollmann, F., Ackermann, S. D. J. U., Bley, T., Walther, T., Steingroewer, J., and Krujatz, F. 2019. Microalgae wastewater treatment: Biological and technological approaches. *Engineering Life Science.* 19:860–871. doi: 10.1002/elsc.201900071.

[51] Grady, L., Daigger, G., Love, N., and Filipe, C. (2011). *Biological wastewater treatment*, 3rd Edition. Boca Raton, FL: CRC Press. doi: 10.1201/b13775.

[52] Chambers, P. A., McGoldrick, D. J., Brua, R. B., Vis, C., Culp, J. M., and Benoy, G. A. 2012. Development of environmental thresholds for nitrogen and phosphorus in streams. *Journal of Environmental Quality.* 41:273. doi: 10.2134/jeq2010.0273.

[53] Hendriks, A. T. W. M., and Langeveld, G. 2017. Rethinking wastewater treatment plant effluent standards: Nutrient reduction or nutrient control. *Environmental Science Technology.* 51:4735–4737. doi: 10.1021/acs.est.7b01186.

[54] Singh, S. P., and Singh, P. 2014. Effect of $CO_2$ concentration on algal growth: A review. *Renewable Sustainable Energy Revision.* 38:172–179. doi: 10.1016/j.rser.2014.05.043.

[55] Singh, S. P., and Singh, P. 2015. Effect of temperature and light on the growth of algae species: A review. *Renewable Sustainable Energy Revision.* 50:431–444. doi: 10.1016/j.rser.2015.05.024.

[56] Almomani, F., Ketife, A. A., Judd, S., Shurair, M., Bhosale, R. R., and Znad, H., et al. 2019. Impact of $CO_2$ concentration and ambient conditions on microalgal growth and nutrient removal from wastewater by a photobioreactor. *Science Total Environment.* 662:662–671. doi: 10.1016/j.scitotenv.2019.01.144.

[57] Ji, M., Abou-Shanab, R., Hwang, J., Timmes, T., Kim, H., and Oh, Y., et al. 2013. Removal of nitrogen and phosphorus from piggery wastewater effluent using the green microalga *Scenedesmus obliquus. Environmental Engineering.* 139:1198–1205. doi: 10.1061/(ASCE)EE.1943-7870.0000726.

[58] Kothari, R., Pathak, V. V., Kumar, V., and Singh, D. P. 2012. Experimental study for growth potential of unicellular alga *Chlorella pyrenoidosa* on dairy wastewater: An integrated approach for treatment and biofuel production. *Bioresource Technology.* 116:466–470. doi: 10.1016/j.biortech.2012.03.121.

[59] Choi, H. J. 2016. Parametric study of brewery wastewater effluent treatment using *Chlorella vulgaris* microalgae. *Environmental Engineering. Research.* 21:401–408. doi: 10.4491/eer.2016.024.

[60]  Gonçalves, A. L., Pires, J. C. M., and Simões, M. 2017. A review on the use of micro-algal consortia for wastewater treatment. *Algal Research*. 24:403–415. doi: 10.1016/j.algal.2016.11.008.

[61]  Al Ketife, A. M., Judd, S., and Znad, H. 2017. Synergistic effects and optimization of nitrogen and phosphorus concentrations on the growth and nutrient uptake of a freshwater *Chlorella vulgaris*. *Environmental Technology*. 38:94–102. doi: 10.1080/09593330.2016.1186227.

[62]  Katam, K., and Bhattacharyya, D. 2021. Improving the performance of activated sludge process with integrated algal biofilm for domestic wastewater treatment: System behavior during the start-up phase. *Bioresource Technology Reports*. 13:100618. doi: 10.1016/j.biteb.2020.100618.

[63]  Christenson, L., and Sims, R. 2011. Production and harvesting of microalgae for wastewater treatment, biofuels, and bioproducts. *Biotechnology Advances*. 29:686–702. doi: 10.1016/j.biotechadv.2011.05.015.

[64]  Ruiz, J., Álvarez-Díaz, P. D., Arbib, Z., Garrido-Pérez, C., Barragán, J., and Perales, J. A. 2013. Performance of a flat panel reactor in the continuous culture of microalgae in urban wastewater: Prediction from a batch experiment. *Bioresource Technology*. 127:456–463. doi: 10.1016/j.biortech.2012.09.103.

[65]  Ahuja, S. 2015. Overview of global water challenges and solutions. *ACS Symposium Series*. 1206:1–25. doi:10.1021/bk-2015-1206.ch001.

[66]  Briggs, A. M., Cross, M. J., Hoy, D. G., Blyth, F. H., Woolf, A. D., & March, L. 2016. Musculoskeletal health conditions represent a global threat to healthy aging: A report for the 2015 world health organization world report on ageing and health. *The Gerontologist*, 56(Suppl 2):243–255. doi: 10.1093/geront/gnw002PMID:26994264.

# 7 A Comprehensive Analysis on Bio-Reactor Design and Assessment Biological Pre-Treatment of Industrial Wastewater

*Santhana Sellamuthu, Zaira Zaman Chowdhury,
Masud Rana, Ahmed Elsayid Ali, Rahman Faizur
Rafique and Seeram Ramakrishna*

## 7.1 INTRODUCTION

Biological remediation of wastewater is a centuries-old biological technique. Until now, since the quantity of discharged industrial wastes is increasing and the constituent of contaminants in effluent solution are diversifying, treatment techniques for wastewater are being intensively researched and tested on a global scale. It is always preferable to combine wastewater remediation with waste utilisation. In this context, this is imperative to suggest as well as make improvements to effluent processing and purification processes in order to increase their overall economy and energy efficiency. It is required to examine the evolution of biological treatment of wastewater techniques and accompanying bioreactor design.

The utilisation of membrane bioreactors (MBR) for such purpose of treating wastewater and reusing it is one potential approach for treating wastewater. Depending on the nutrient removal goals of a particular project, a membrane bioreactor (MBR) can operate under an aerobic or anaerobic mode. Aerobic MBRs are typically applied for the treatment of wastewater generated from residential sources, "night soil," as well as industrial and municipality water. The majority of anaerobic MBR applications have been made for industrial wastewaters with a high organic concentration. Because anaerobic bacteria exhibit lower rate of growth than aerobic bacteria, they generate less leftover sludge but require a proportionally longer retaining time. This is because anaerobic bacteria need the presence of oxygen throughout their growth. In addition, because of their fluid and fibrous properties, anaerobic bio-solids have a poor settlement rate. As a result, anaerobic MBRs offer benefits over traditional methods. MBR techniques that use aerobic-anoxic cycling to achieve optimum de-nitrification have been implemented in situations when it is necessary to

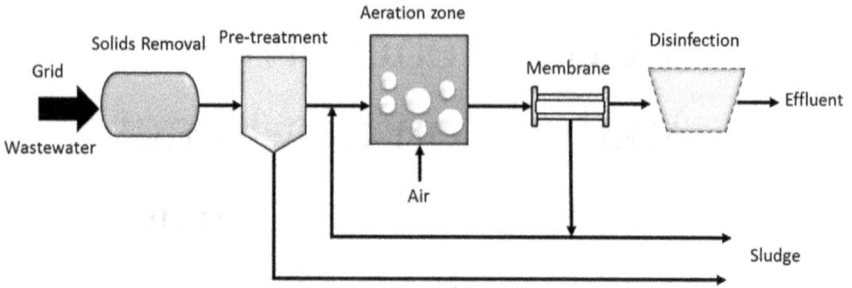

**FIGURE 7.1**   Process of Membrane Bioreactor (MBR).

completely remove nitrogen [1–3]. Uses of MBR have included filtering of groundwater and oily pollutants, landfill leachate, pharmaceutical byproducts, phosphorus, and chlorinated solvents in industrial plant wastewaters. Figure 7.1 illustrates the basic principle of MBR operation for wastewater treatment.

Recent research has led to the invention of a method for the treatment of wastewater, known as the membrane bioreactor (MBR) system [4, 5]. This method consists of two stages: a biological phase and a membrane module. Without the traditional setup of an aeration tank, settling tank, and filtration, it is possible to treat wastewater to obtain high quality and generate tertiary grade effluent with the ratio of 5:5:5 for suspended particles, biological oxygen demand (BOD), and ammonia. Because of this and a number of other advantages, the MBR system is now specifically tailored for the remediation of toxic wastewater from industrial processes as well as the reclamation of pure water [6, 7]. The flow is allowed to pass over the membrane, but the solids are kept within the biological treatment systems. The advantages of the suspending growth bioreactors are combined with the capacity for solid separation offered by either an ultra-filter or a micro-filter membrane device inside the membrane bioreactor system. Because membranes with pore sizes that are typically in the range of 0.1–0.5 micrometres retain a considerable amount of harmful microorganisms, the need for disinfection is also decreased. Because the membrane has a prolonged retention time for solids – typically 30–60 days – this feature can significantly boost the biological decomposition of organic substances that are present in the feed water. Typical values for MLSS in submerged MBRs vary from 12–15 g/L to 30 g/L, while tubular systems for the treatment of industrial wastewater can achieve up to 30 g/L. Figure 7.2 depicts the schematic of the membrane bioreactor treatment process using a variety of configurations.

The primary advantage of this method is its compactness, since the clarifier, where sludge is normally separated from treated effluent by gravitational force, is replaced with a filtration membrane that may be employed immediately in the biological aerobic reactor. In addition, the membrane unit can function with a sludge having concentration up to 20–25g total solids (TS)/L in the bioreactor. This is in contrast to the traditional technology, which is restricted to a maximum of 5g/L to maintain the efficient sedimentation of sludge. In addition, unlike conventional technology, MBR plants may be run under a wider range of operating circumstances, including

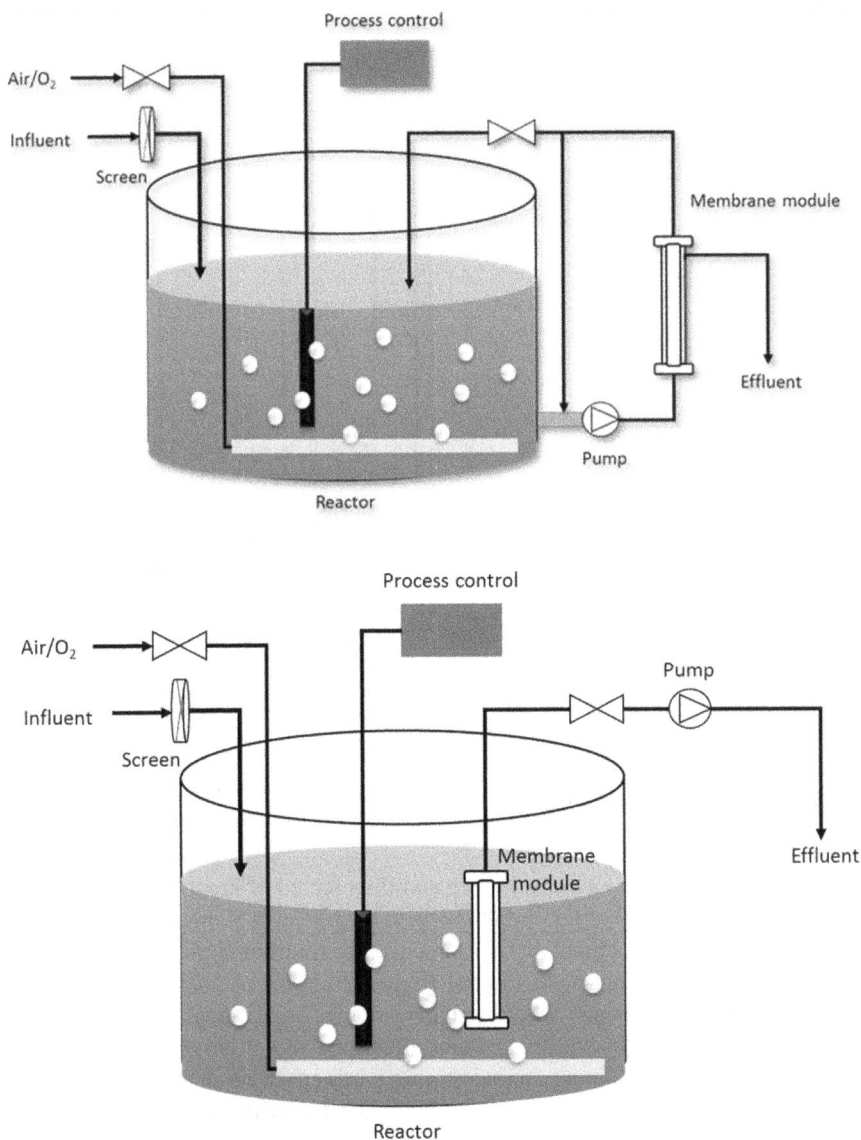

**FIGURE 7.2**   Basic Layout of Membrane Bioreactor (MBR) (a) External and (b) Submerged.

concentration and age of sludge, organic feed load, etc. and have greater resilience
to pressure changes [7, 8]. Furthermore, the adaptability of the technique makes it
simpler to implement and plan in regions experiencing rapid population increase,
where it might be difficult to estimate the volume of water that will need to be purified beyond the next upcoming years. Finally, the membrane bioreactor (MBR) technology stands out because of the outstanding and reliable treatment superiority that
can be accomplished with it. Figure 7.3 illustrates the basic layout of MBR set-up.

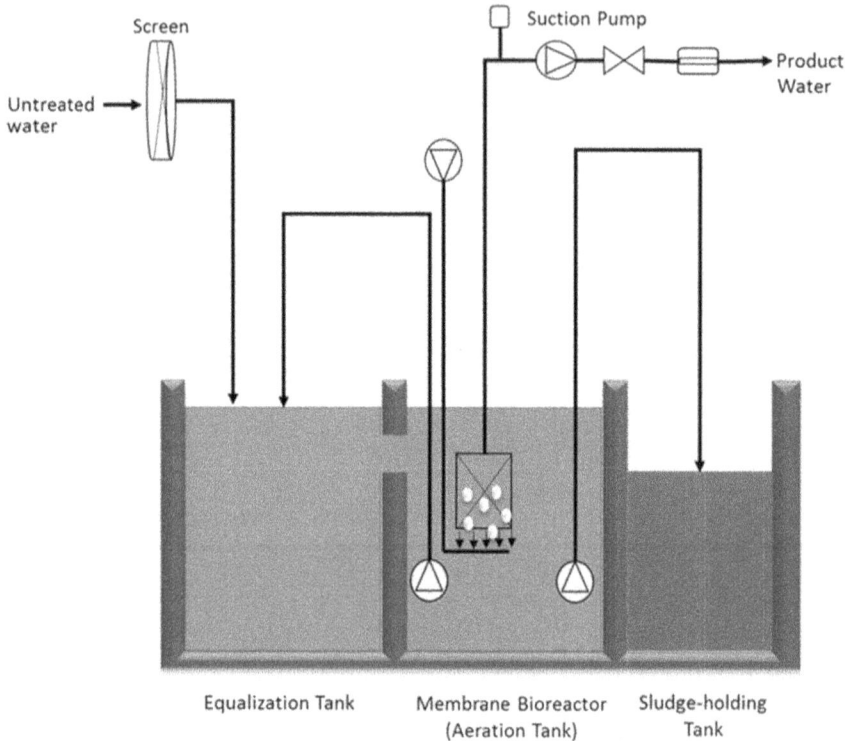

**FIGURE 7.3**   Basic Layout of Membrane Bioreactor (MBR).

This includes producing particle-free and disinfected effluent regardless of the amount of pollutant load or incoming raw water, while also being able to overcome typical fundamental problems that are found using traditional plants, such as bacterial populations, bulking or floating of sludge, presence of flocs inside pinpoint, and vice versa. In situations where stringent processing standards are necessary, such as compliance with swimming water directions and/or unlimited recycling of water, the MBR treated water becomes highly pertinent because of this. As a result of the superior quality of the permeate generated by MBR, which is free of not only particles and bacteria but also colloids, the MBR technique is also an effective pre-treatment prior to the implementation of nano-filtration (NF) or reverse osmosis (RO).

## 7.2   HISTORY FOR DEVELOPMENT OF MBR TECHNOLOGY

The application of MBRs for treating industrial and municipal wastewater has witnessed substantial growth during the past decade. This is primarily owing to its functionality of removing inorganic and organic pollutants along with microorganisms from effluents. In recent decades, it has managed to gain wide acceptance as a direct consequence of stricter environmental laws and growing approaches to recycle water.

Independent groups working in Japan and Canada in the 1980s or early 1990s came up with the idea of submerged or immersed membranes. These membranes might be used for a variety of applications. Research from Rensselaer Polytechnic Institute in Troy, New York and Dorr-Oliver, Inc. in Milford, Connecticut, United States [9] were the first to describe the concept of linking the activated sludge process with membrane separation.

In spite of the fact that it did not attract much attention in North America, it had a great deal of popularity in Japan during the 1970s and 1980s. In 1989, when membranes were first submerged in the bioreactor, a significant step forward was taken for the MBR technology. In 1989, the government of Japan initiated a research and development project that was to last for one year. The goal of the project was to develop low-cost treatment processes that made use of MBR in order to generate recyclable water from wastewater. This program contributed to the development of various technologies, such as the flat-sheet-based module developed by Kubota and Hitachi-Plant and the hollow-fiber-based module developed by Zenon and Mitsubishi Rayon [9].

## 7.3   TYPES OF CONFIGURATION/DESIGN IN MBR

The membrane microarray can be configured in two different ways: either the membranes will be placed outside the bioreactor or they will be positioned inside [10, 11]. A high water velocity across the filter channel helps to prevent fouling, as demonstrated in Figure 7.2 for the exterior arrangements. This is accomplished by moving water quickly through the channel. The pressure that is generated as a result of a fast cross-flow moving in the direction of the membrane is the major driver. This structure, as a consequence, allows more direct dynamic controlling of the membrane fouling. It also gives the benefits of faster replacement of membrane with higher water fluxes, but at the disadvantage of periodic cleansing as well as a high level of energy usage (2–12 kWh/m$^3$ products). In designs that use submersion, membrane modules are embedded straight into the liquid that is being mixed [10, 11]. The driving force that is needed to move the permeate water across the membrane can be generated by either pressurizing the bioreactor or establishing a pressure drop inside the membrane reactor that retains the permeate water [10–12]. Both the pressure variations across the trans-membrane and the flux rates are quite minimal. This particular type of membrane is utilized for the treatment of wastewater in municipal and industrial settings. It has the flexibility to be installed either inside the external filtration tank or within the aeration tank. In comparison to the lateral stream system, the operational flux of the membrane requires cleaning that is less regular and less stringently done in order to be restored.

## 7.4   TYPES OF MEMBRANE BIOREACTOR

### 7.4.1   MBR with Aerobic Stirred Reactor

One of the oldest commercial processes of biotechnology is the activated sludge technique, which features the remediation of industrial waste stream in bioreactors under aerobic condition using a stirred tank. The standard activated sludge method utilizes a stirred tank with agitation and a separate tank, under aerobic condition which is inoculated with a microbiological sludge inoculum (generally the reprocessed fraction of active sludge). At this stage, suspended microorganism growth occurs. For the purpose of providing adequate dissolved oxygen (DO) in a media, air is sprayed from the bottom under high pressure. Subsequently the volume generated by the aerobic tank is typically quite massive and the dissolution rate of atmospheric $O_2$ in water or aqueous effluent is very minimal; enormous air compressors would be required to spurge in a substantial volume of air in order to satisfy the $O_2$ requirements of the microbial species and initiate the process under the aerobic condition. Even though the technology is simple to develop and install, the increasing operating expenditure of air compressors is the primary financial constraint of this method.

At this stage, the oxidation of dissolved organic constituents takes place to generate $CO_2$ and $H_2O$, and nitrification and denitrification could also occur at this point. Due to nitrification, dissolved ammonia in wastewater is transformed to nitrates within the aerobic tank itself (concurrently with removal of carbon). As an anaerobic process, denitrification is undertaken in a different bioreactor. At the phase of denitrification, the nitrates produced by nitrification are converted into nitrogen gas, which is then evacuated from the bioreactor. As the operation is controlled under anaerobic condition, this bioreactor does not require an external stream of air and has provided a thorough examination of the subject [13].

The traditional system consists of two bioreactors (stirred tanks) in sequence, the first of which is an aerobic tank where the elimination of carbon (organic matter destruction) and nitrification takes place, the second of which is an anaerobic tank in which denitrification takes place. The discharge from the denitrification tank is transferred to the unit of sedimentation for clarity; cleaned water spills with the thicker sludge from the bottommost part and is largely recirculated inside the tank operated under aerobic condition (stirred tank-1). The aggregate of microbial recycled sludge must be regulated to decrease the rate of spontaneous deterioration of bacteria while retaining a greater percentage of biological oxygen demand (BOD) elimination.

Moreover, it is feasible to employ a series-parallel configuration of agitated tanks to further improve the efficiency of the system (tank under aerobic condition) [14–16]. In this instance, too, the cell is separated into a multitude of units under aerobic condition, with each receiving a portion of the unprocessed polluted water (effluent feed) and being aerated independently. Therefore, both step feeding and step aeration are applied. Each compartment, with the exception of the first, obtains a proportion of the incoming feed along with the incompletely treated wastewater from the unit preceding it. This strategy is suitable for installations with a high capacity. Each compartment might function similarly to an optimal continuous stirring tank reactor (CSTR) in this case, allowing for close substrate-biocatalyst interaction (microbial cells).

It is reasonable (and frequently recommended) to pair membrane based devices with activated sludge processing. After pre-treatments (such as lime inclusion, coagulation, filtration, and clarification), the wastewater is fed to the RO unit, from which the permeate water is recovered. In the aerobic tank and denitrification bioreactor, the RO concentrate undergoes additional biological treatment. In this regard, one of the research groups [17] has published a successful prosecution study and proved that the BOD, nitrogen, and phosphate exclusion can be amplified by combining reverse osmosis (RO) equipped under aerobic conditions. It has been reported in an economic analysis of this method [18]. The operational pressure of the RO plant (the inter-membrane pressure difference that must be sustained) and the usable life expectancy of the polymeric membrane have the greatest impact on the entire economics of the RO system. Possibilities of membrane blockage and fouling are further complications. According to the researchers, two-thirds of the waste effluents could be retrieved in the RO unit and the remaining one-third of the total waste could be submitted to biological treatment, hence reducing the total cost of producing treated water by three-quarters compared to the conventional system [18]. This includes the expense of membrane replacement. Based on their laboratory research, the research team [12] has demonstrated that deploying an up-stream forward osmosis (FO) unit and a down-stream nano-filtration unit would assist in achieving a high extent of chemical oxygen demand (COD) elimination (> 97%) from pharmacological waste effluents. However, the overall economics of the plan must be examined with the high cost of operation of the nano-filters and the vast quantity of industrial effluents that must be treated in mind.

### 7.4.2 MBR WITH ANAEROBIC STIRRED REACTOR

For anaerobic treatment of commercial, home, and municipal wastewater, bioreactors with stirred tank are among the oldest and most widely used technologies, similar to the scenario with aerobic waste treatment. Greater capacity (large holdup) and simplicity of deployment are the primary factors for this decision. Treatment of wastes/effluents under anaerobic condition has the added benefit of converting organic matter into high-value yields such as biogas (which is primarily a combination of $CH_4$ and $CO_2$) with a complex mixture of acidogenic, acidogenic, and methanogen microbes. The anaerobic digested sludge might be employed easily as an inferior grade nitrogenous bio-fertilizer or utilised in the biochemical route [10, 11] for the production of phosphorus bio-fertilizers (called phosphate rich organic manure).

However, the anaerobic digesting process is somewhat slower. Also, as obligatory organisms, methanogen bacteria are particularly sensitive to the working temperature of T = 330–35 °C and pH 7.0. Using thermophilic microorganisms, anaerobic digestion at higher temperatures (55–65°C) is also possible. This expedites the destruction of germs, but there will be additional costs associated with the construction of heating pipes and the external supply of heat. Frequently, the expense of additional energy input tends to outweigh the advantages of faster microbial elimination and greater methane output. In addition, thermophilic microbes have a slower growth rate than mesophilic bacteria. Unless waste heat is accessible, such as in combined power and heat systems, thermophilic waste treatment shall not be desirable or advantageous. Nevertheless, a thermophilic pre-treatment may be administered to the feed slurry if pathogen killing is a major issue [3, 12]. Paper waste is co-digested with cow manure and water hyacinth to achieve higher biogas production. It has been shown that the mixing of microalgae (algal mass) boosts the rate of anaerobic digestion of peat hydrolysate, domestic sewage sludge, and waste sulphite liquor, resulting in a greater biogas output [14]. Research on the co-digestion of micro-algae with urban food wastes is described by other researchers [18]. In every instance, the addition of algal material was found to increase the speed of digesting and the production of biogas [16]. According to other researchers [19–21], anaerobic co-digestion of several substrates, like sewage sludge and paper waste, with waste-grown algae results in a much greater biogas output. When utilized as a sewage sludge, specific substrate, waste paper, and algal mass produced an average of 275, 120 and 200 mL d$^{-1}$, but when all three substrates were co-digested, the biogas yield jumped to between 550 and 600 mL/d $^{-1}$.

In stirred tank bioreactors, stirring of the slurry materials is one of the working challenges encountered. Introduction of a mechanical impeller is troublesome because it could result in ambient air leaking into the bioreactor and biogas escaping. Occasionally, it is advantageous to divide the reactor into two chambers using a partially sunken partition, one of the early methods. The feed material runs below the initial section, over the divider, inside the second partition, and then into the outflow tank as digested sludge (sludge tank). This winding movement of the slurry persuades the turbulence and mixing inside the slurry. Then drifting gas container is utilized; its movement in longitudinal direction generates some basal agitation [22].

### 7.4.3 MBR WITH FLUIDIZED BED REACTOR

Fluidized bed biofilm reactors are the bioreactors of choice for higher capacity installation because these types of bioreactors may be operated at significantly higher speed of fluid and feed flow rate [23]. The effluent from industries is allowed to enter at the lowermost part of the column at a speed that is significantly faster than the speed of the feed flow but slower than the final independent velocity gradient of each particle-biofilm aggregation. As a direct consequence of this, all particles continue to exist in the rising torrent of reaction mixture in a fluidized state.

Because there is no channeling, each accumulation is completely surrounded by the particulate. As a result, the interacting between the two is more immediate, which contributes to an increase in the efficiency of the bioreactor. In addition to this, the total active volume of the reactor expands as the bed keeps expanding (the height of the extended bed, $L_f$, is determined by the operational flow rate of fluid that is being used).

Once the bed has been completely fluidized, the pressure drop across the bed doesn't really increase with an increase in the fluid rate, which is another unique aspect of these bioreactors. Once the bed has been entirely fluidized, drop of pressure throughout the bed stays more or less persistent. Because of this, there is not going to be a significant difference in the overhead expenses of the bioreactor even if the feed flow rate is increased. Researchers [24] looked into the possibility of using aerobic processes for treating wastewater in three-phase fluidized bed biofilm reactors. Their investigation includes a comprehensive mathematical modelling and simulation of the performance of bioreactors, followed by a verification of the simulated data by a comparison with in-depth experimental data gathered both on a lab scale and on a pilot scale plant.

Several authors have written about their experiences conducting lab-scale research on fluidized bed bioreactors. For instance, researchers [25, 26] investigated the de-phenolization of waste stream in a lab setting fluidized column constituted of immobilized *Pseudomonas cells*. Laboratory studies on microbial mercury separation in a three-phase fluidized bed was reported [27]. Both of these studies were conducted in laboratories. Studies on the production of anaerobic lactic acid from sugar mills and dairy waste (cheese whey) wastewaters (molasses) in a fluidized bed biofilm reactor have been presented by earlier researchers. They used *Lactobacillus helveticus* culture for the earlier and *Enterococcus faecalis* culture for the latter one [28, 29].

### 7.4.4 MBR WITH SEMI-FLUIDIZED BED REACTOR

Within the industry of wastewater treatment, semi-fluidized bed biofilm reactors are one of the more recent developments. Even though semi-fluidization technologies have greater running costs, they offer a number of benefits that cannot be replicated by traditional fluidized bed technologies. Because the fluid flow rate that is used in semi-fluidized bed bioreactors is greater than that which is implemented in traditional fluidized bed bioreactors, these reactors are capable of functioning at larger capacities.

The researchers have presented their research on the aerobic purification of industrial waste discharge for the elimination of BOD and the degradation of o-cresol in three-phase semi-fluidized bed bioreactors [30]. Earlier research was conducted on the aerobic treatment of petrochemicals and distillery effluent in a semi-fluidized bed based biofilm reactor [31]. The process of the bioreactor is further reliable, and it has been seen to provide a higher degree of BOD/COD reduction at higher flow rates of feed, despite the demand for a relatively minimal amount of reactor volume. The elimination of the need for air compressors resulted in a significant reduction in the running costs associated with the reactor. Both mathematical analysis and practical verification support the findings of the observations. The layout of semi-fluidized bed reactor is shown by Figure 7.4.

### 7.4.5 MBR WITH DSFF REACTOR

In contrast to the traditional biofilm reactors that were described earlier, DSFF bioreactors do not make utilization of any support particles. This is the primary distinction between the two types of bioreactors. The formation of the biofilm takes place on the inside wall or exterior of one or even more vertically up-righted flow channels or pipes, which are the conduits through which the feeding solution travels (Figure 7.5). As a result of their down-flow style of operation, they are not only inexpensive to run but also quite straightforward in terms of construction and upkeep.

DSFF bioreactors were used for the biological treatment of wastewater under anaerobic conditions [31, 321]. In their study on the anaerobic degradation of brewery wastewater, the researchers have [33] associated the enactment of a DSFF bioreactor with that of an up-flow anaerobic sludge-blanket (UASB) bioreactor, a fluidized bed biofilm reactor, and an up-flow anaerobic filter (UAF). All three bioreactors were up-flow designs. According to their findings, the DSFF and UASB bioreactors both remove a similar percentage of chemical oxygen demand (COD) from the

Gases exhaust

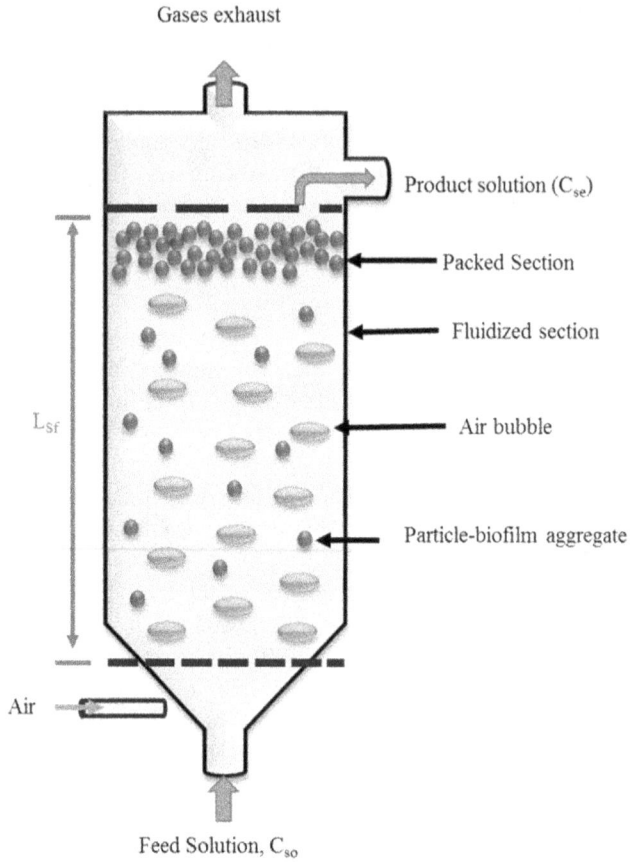

Product solution ($C_{se}$)

Packed Section

Fluidized section

$L_{sf}$

Air bubble

Particle-biofilm aggregate

Air

Feed Solution, $C_{so}$

**FIGURE 7.4**   Layout of Semi-Fluidized Bed Reactor.

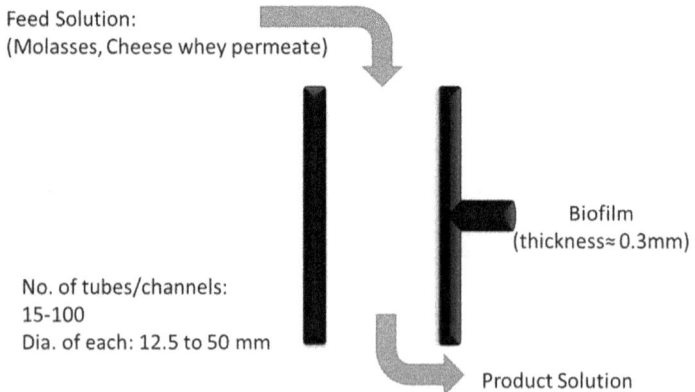

Feed Solution:
(Molasses, Cheese whey permeate)

Biofilm
(thickness≈ 0.3mm)

No. of tubes/channels:
15-100
Dia. of each: 12.5 to 50 mm

Product Solution

**FIGURE 7.5**   Basic Layout of DSFF Reactor.

water (77%) whereas the UAF and fluidized bed biofilm reactor have the potential to remove more than 90% of both BOD and COD.

## 7.4.6 MBR WITH INVERSE FLUIDIZATION BED REACTOR

In case of "inverse fluidization," the feed solution is introduced into these bioreactors using a down-flow mechanism. Because the density of the particle-biofilm aggregation (as specified) is lower than that of the reaction mixture, they continue to float in the descending stream of the mixture. This is because the concentration of the aggregates is lesser than that of the working solution (Figure 7.6). The beads of polymer, which have a density lower than that of water, are frequently used in these bioreactors in the capacity of support particles. The down-flow mode of operation of these bioreactors, which removes the need for costly pumping of feed solution, is the primary contributor to their advantageously low operating costs. This is the most substantial advantage of this type of reactor.

Numerous writers [34–41] have published the results of experimental research on the aerobic treatment of wastewater carried out in laboratory inverted fluidized bed biofilm reactors. Aerobic processing of wastewater generated from refineries (for de-phenolization) and brewing waste effluents [34–37], starch effluent [38, 39], and municipal wastewater specimens [40] are some instances of this type of treatment. Despite being conducted on a laboratory bench scale, the experimental data that were presented by these researchers are promising. It has been claimed by researchers [41] that oil can be removed from wastewater by using an inverse fluidized bed column that is made up of aerogel granules. It has been stated that approximately 95% of the oil may be recovered from water that has been contaminated.

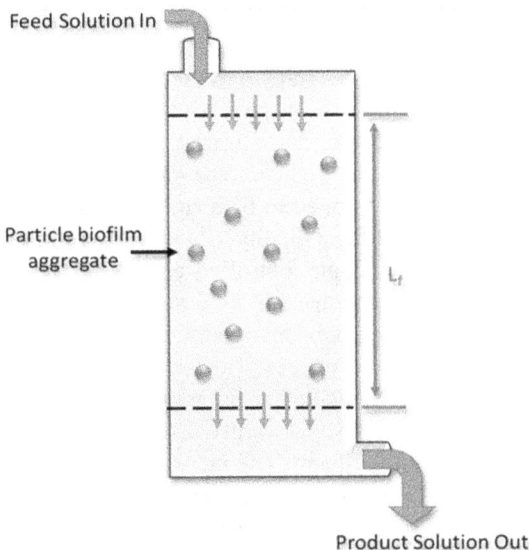

FIGURE 7.6   Layout of Inverse Fluidized Bed Reactor.

## 7.5 CONCLUSION

The effluent from industrial wastewater is evidently filled with hazardous heavy metals, dyes, and other organic compounds, all of which are unlikely to be processed adequately and, as a consequence, contribute to the pollution of the environment. The degradation of the ecosystem that results from the discharge of effluents from industrial processes that are not adequately handled leads to the unavailability of the pure water supply to meet the essential needs of life. It is believed that effluents emitted by industries such as paper, textile, pulp and oil, brewery, food and beverage, etc. can produce large quantity of industrial wastewater as well as the challenges associated with disposal. This chapter is a review of possible courses of action that could be implemented in the treatment process for the reuse and recovery of water using MBR. Additionally, this chapter inspects the problems that may arise in relation to treatment systems as well as the technical solutions that may be available to solve those problems. Although there have been developed a variety of traditional and conventional treatment methods for the purpose of treating industrial wastewater, the execution of integrated water reuse design enhances the use of recovered wastewater that provides sufficient operability to ensure adherence in water supply. This is accomplished through the utilisation of reclaimed wastewater.

The predominant grade of this is based on the present regulatory environment for wastewater management, and it is recognized to support prospective research efforts in this sector. This quality was determined by taking into account the current environment. The literature is reviewed in such a way that the focus of the paper is on techniques and their applicability in industries, as well as the efficiency, design considerations, and selectivity of numerous industrial effluent management plants based on MBR.

In the upcoming future, membrane bioreactors (also known as MBRs) will be employed wherever high standard effluents are necessary, whether it be because of a delicate receiving reservoir or because of the fact that water is reused as treated water. When additional management coupled with a nano-filtration (NF) or reverse osmosis (RO) is being explored for an industrial application, MBRs are the ideal choice for the pre-treatment stage. The combination of activated sludge with membrane filtration has been demonstrated to be a straightforward, one-step procedure that can result in exceptional effluent quality.

Recent advancements in membrane control system have led to relative profitability for exterior versus interior membrane MBRs throughout a much greater range of wastewater flow rates. This was previously only possible with internal membrane MBRs. Future advancements are expected to involve the advent of cost-efficient anaerobic MBR systems and the implementation on a large scale of alternate MBR arrangements where the membranes are utilized for reasons rather than the separation of biomass based waste effluent. The increasing acceptability of MBRs, the rising attention in water reclamation and reprocessing technologies, and the rising need for progressive solutions for purification of wastewater have led to a prosperous future for MBR technologies.

## ACKNOWLEDGMENTS

The authors are thankful for the funding provided by Malaysian Joint Research Scheme-ST 077–2022, Interdisciplinary Research IIRG003A-2022IISS and International Grant ICF 023–2022 and ICF 080–2021 under University of Malaya, Kuala Lumpur 50603, Malaysia.

## REFERENCES

[1]   Marrot, B., Barrios-Martinez, A., Moulin, P., and Roche, N. 2004. Industrial wastewater treatment in a membrane bioreactor: a review. *Environmental Progress*, 23(1), pp. 59–68.

[2]   Adebayo, G.B., Kolawole, O.M., Ajijolakewu, A.K., and Abdulrahaman, S.O. 2010. Assessment and biological treatment of effluent from a pharmaceutical industry. *Annals of Biological Research*, 1(4), pp. 28–33.

[3]   Agboola, O., Babatund, D.E., Fayomi, O.S., Sadiku, E.R., Popoola, P., Moropeng, L, Yahaya, A., and Mamudu, O.A. 2020. A review on the impact of mining operation: monitoring assessment and management. *Results in Engineering*, 8, p. 100181.

[4]   Ahmad, T., Aadil, R.M., Ahmed, H., Rahman, U., Soares, B.C., and Souza, S.L. 2019. Treatment and utilization of dairy industrial waste: a review. *Trends Food Science Technology*, 88, pp. 361–372.

[5]   Lin, H., Gao, W., Meng, F., Liao, B.Q., Leung, K.T., Zhao, L., and Hong, H. 2012. Membrane bioreactors for industrial wastewater treatment: a critical review. *Critical Reviews in Environmental Science and Technology*, 42(7), pp. 677–740.

[6]   Anderson, G.K., Kasapgil, B., and Ince, O. 1996. Microbial kinetics of a membrane anaerobic reactor system. *Environmental Technology*, 17(5), pp. 449–464.

[7]   Sutton, P.M. 2006. Membrane bioreactors for industrial wastewater treatment: applicability and selection of optimal system configuration. *Proceedings of the Water Environment Federation*, (9), pp. 3233–3248.

[8]   Chang, I.S., Le Clech, P., Jefferson, B., and Judd, S. 2002. Membrane fouling in membrane bioreactors for wastewater treatment. *Journal of Environmental Engineering*, 128(11), pp. 1018–1029.

[9]   Radjenovic, J., Matošic, M., Mijatovic, I., Petrovic, M., and Barceló, D. 2008. Membrane bioreactor (MBR) as an advanced wastewater treatment technology. In *Emerging Contaminants from Industrial and Municipal Waste*, pp. 37–101. Berlin, Heidelberg: Springer.

[10]  Sutton, P.M. 2006. Membrane bioreactors for industrial wastewater treatment: applicability and selection of optimal system configuration. *Proceedings of the Water Environment Federation*, (9), pp. 3233–3248.

[11]  Abdel-Kader, A.M. 2007. *A review of Membrane Bioreactor (MBR) technology and their applications in the wastewater treatment systems*. In Eleventh International Water Technology Conference, IWTC11 Sharm El-Sheikh, Egypt, pp. 269–278.

[12]  Thakura, R, Chakraborty, S., and Pal, P. 2015. Treating complex industrial waste water in a new membrane—integrated closed loop system for recovery and reuse. *Clean Technology and Environmental Policy*, 17, pp. 2299–310.

[13]  Rao, K.R., and Subrahmanyam, N. 2004. Process variations in activated sludge process—a review. *Indian Chemical Engineering*, 46, pp. 48–55.

[14]  Zabot, G.L., Mecca, J., Mesomo, M., Silva, M.F., Pra, V.D., Oliveira, D., et al. 2011. Hybrid modeling of xanthan gum bio-production in batch bioreactor. *Bioprocess Biosystem Engineering*, 34, pp. 975–86.

[15]   Narayanan, C.M. 2012. Production of phosphate rich biofertiliser using vermicompost and anaerobic digester sludge—a case study. *Advance Chemical Engineering Science.* 2, pp. 187–91.

[16]   Olsson, J., Feng, X.M., Ascue, J., Gentili, F.G., Shabiimam, M.A., Nehrenheim, E., et al. 2014. Co-digestion of cultivated microalgae and sewage sludge from municipal waste water treatment. *Bioresource Technology*, 171, pp. 203–10.

[17]   Smith, R. 1970. *USEPA report: 170-40-05-70.* Washington, DC: US Environmental Protection Agency.

[18]   Narayanan, C.M. 1993. Energy conservation employing membrane-based technology. *Chemical Industry Digest*, 6, pp. 133–6.

[19]   Ajeej, A., Thanikal, J.V., Narayanan, C.M., and Kumar, R.S. 2015. An overview of bio augmentation of methane by anaerobic co-digestion of municipal sludge along with microalgae and waste paper. *Renewable Sustainable Energy Revision*, 50, pp. 270–276.

[20]   Ajeej, A., Thanikal, J.V., Narayanan, C.M., and Yazidi, H. 2016. Studies on the influence of the characteristics of substrates in biogas production. *International Journal of Current Research*, 8, pp. 39795–9.

[21]   Ajeej, A., Thanikal, J.V., and Narayanan, C.M. 2016. Studies on production of biogas by codigestion of sewage sludge, wastepaper and waste grown algae. *Journal of Modern Chemistry and Chemical Technology*, 7, pp. 74–81.

[22]   Rahman, M.H., and Al-Muyeed, A. 2010. *Solid and hazardous waste management.* Dhaka: Center for Water Supply and Waste Management.

[23]   Shieh, W.K., and Keenan, J.D. 1986. Fluidized bed biofilm reactor for wastewater treatment. In *Bioproducts*. Advances in Biochemical Engineering/Biotechnology, vol. 33, pp. 131–69. Berlin: Springer.

[24]   Narayanan, C.M., and Biswas, S. 2015. Computer aided design and analysis of three phase fluidized bed biofilm reactors for waste water treatment. *Asian Journal of Biochemical and Pharmaceutical Research*, 5, pp. 224–49.

[25]   Gonzalez, G., Herrera, M.G., Garcia, M.T., and Pena, M.M. 2001. Biodegradation of phenol in a continuous process: comparative study of stirred tank and fluidized-bed bioreactors. *Bioresource Technology*, 76, pp. 245–51.

[26]   Gonzalez, G., Herrera, G., Garcıa, M.T., and Pena, M. 2001. Biodegradation of phenolic industrial wastewater in a fluidized bed bioreactor with immobilized cells of Pseudomonas putida. *Bioresource Technology*, 80, pp. 137–42.

[27]   Deckwer, W.D., Becker, F.U., Ledakowicz, S., and Wagner-Dobler, I. 2004. Microbial removal of ionic mercury in a three-phase fluidized bed reactor. *Environmental Science Technology*, 38, pp. 1858–65.

[28]   Narayanan, C.M. 2015. Case studies on synthesis of PLLA bioplastic starting from food and agricultural wastes. *International Journal of Chemical Engineering Processing*, 1, pp. 1–13.

[29.   Narayanan, C.M., Das, S., and Pandey, A. 2017. Food waste utilization: green technologies for manufacture of valuable products from food wastes and agricultural wastes. In *Handbook of food bioengineering*, vol. 2, pp. 1–54. London: Academic Press.

[30]   Narayanan, C.M., and Biswas, S. 2016. Studies on waste water treatment in three phase semifluidized bed bioreactors—computer aided analysis and software development. *Journal of Modern Chemical Technology*, 7, pp. 1–21.

[31]   Henze, M., and Harremoës, P. 1983. Anaerobic treatment of wastewater in fixed film reactors—a literature review. *Water Science Technology*, 15, pp. 1–101.

[32]   Samson, R., Van den Berg, L., and Kennedy, K.J. 1985. Mixing characteristics and start-up of anaerobic downflow stationary fixed film (DSFF) reactors. *Biotechnology Bioengineering*, 27, pp. 10–9.

[33] Jovanovic, M., Murphy, K.L., and Hall, E.R. 1986. *Parallel evaluation of high rate anaerobic treatment processes: retention time and concentration effects*. In EWPCA Conference on Anaerobic Treatment: A Grown-up Technology, Amsterdam.

[34] Sokol, W. 2003. Treatment of refinery wastewater in a three-phase fluidised bed bio-reactor with a low density biomass support. *Biochemical Engineering Journal*, 15, pp. 1–10.

[35] Sokol, W., and Korpal, W. 2005. Phenolic wastewater treatment in a three-phase fluidised bed bioreactor containing low density particles. *Journal of Chemical Technology Biotechnology*, 80, pp. 884–91.

[36] Sokol, W., and Korpal, W. 2006. Aerobic treatment of wastewaters in the inverse fluidised bed biofilm reactor. *Chemical Engineering Journal*, 118, pp. 199–205.

[37] Sokół, W, and Woldeyes, B. 2011. Evaluation of the inverse fluidized bed biologica reactor for treating high-strength industrial wastewaters. *Advance Chemical Engineering Science*, 1, pp. 239–44.

[38] Rajasimman, M., and Karthikeyan, C. 2007. Aerobic digestion of starch wastewater in a fluidized bed bioreactor with low density biomass support. *Journal of Hazardous Materials*, 143, pp. 82–6.

[39] Rajasimman, M., and Karthikeyan, C. 2009. Optimization studies in an inverse fluidized bed bioreactor for starch wastewater treatment. *International Journal of Environmental Research*, 3, pp. 569–74.

[40] Haribabu, K., and Sivasubramanian, V. 2016. Biodegradation of organic content in wastewater in fluidized bed bioreactor using low-density biosupport. *Desalination Water Treatment*, 57, pp. 4322–7.

[41] Quevedo, J.A., Patel, G., and Pfeffer, R. 2009. Removal of oil from water by inverse fluidization of aerogels. *Indian Engineering Chemistry Research*, 48, pp. 191–201.

# 8 De-Ballasting Water Treatment Technology to Protect the Sea-Ocean Environment from Ships at the Western Australian Ports

*Tien Anh Tran*

## 8.1 INTRODUCTION

The maritime transportation industry plays an important role in developing the national economics through the sea-ocean pathway. The international maritime organization (IMO) had given a lot of stringent conventions to ensure the safety and reliability of the seafarers, the equipment on ships including STCW revised 2010, MARPOL 73/78, SOLAS 74, MLC, and so on. The international convention of ballast water management (BWM) is an official rule as well as the international conventions which relate directly to the wastewater treatment management from ships. Currently, there are about 10 billion tons of ballast water due to maritime transportation discharged into the ocean sea environment. There are about 10,000 living species in the ballast water which come into the various maritime ecosystems per day from ships. These harmful species are a main cause of the sea pollution for national economics, the human health system, and the plant environment [1–6]. Then, the ballast water management has received a lot of research as well as the associated national legislation in the United States of America. It is a main reason that the installation of ballast water treatment systems is a mandatory requirement for all existing ships nowadays [7, 8]. The used ballast water could be fresh/salt water depending on the operational route of ships [9]. The management of energy efficiency for existing ships and wastewater discharging is crucial for the stakeholders of shipping transportation companies in the world nowadays. These research results related to the energy efficiency management have been conducted and published through different methodologies [10–13]. Additionally, the wastewater discharging from a ship is a difficult work in managing and controlling the harmful species and pollutant particles. This one is a complicated problem related to the marine policies and environmental management. So, the development of the guidelines and technical innovation makes

DOI: 10.1201/9781003342830-8

it very difficult to determine the wastewater discharge from ships (ballast water) according to the standard convention [14] (Figure 8.1).

The importance of this wastewater management has been determined through there are some different methodologies proposed to treat completely before discharging into the ocean sea environment. The ballast water treatment mechanism includes three types of treatment process such as the mechanical, physical, and chemical one [15] like Figure 8.2.

In reality, the ballast water management is an important system in response to the stringent regulations of the international maritime organization (IMO). The appropriate procedures of this system will be studied and investigated in this research. A certain ship will be selected to validate and utilize the ballast water management under the guidance from the international regulations due to IMO approval. The Western

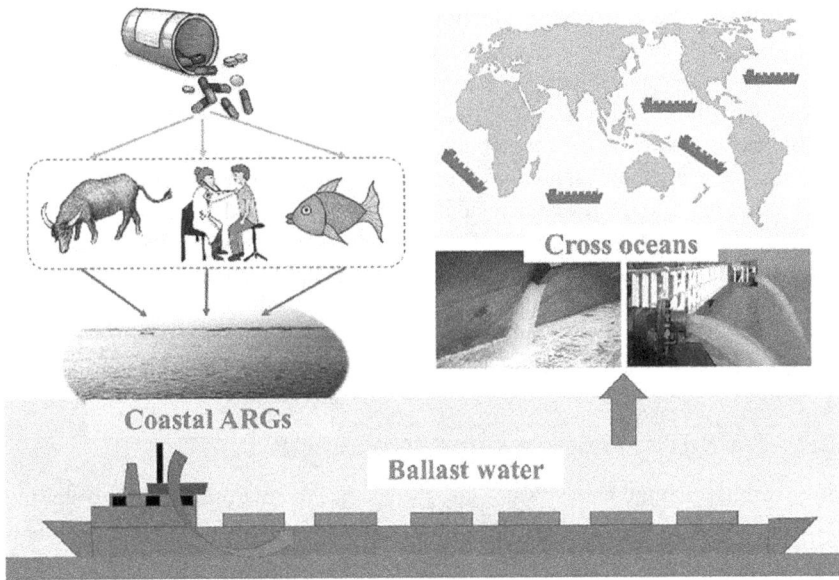

**FIGURE 8.1** The critical issues to manage the wastewater discharging from ship (ballast water).

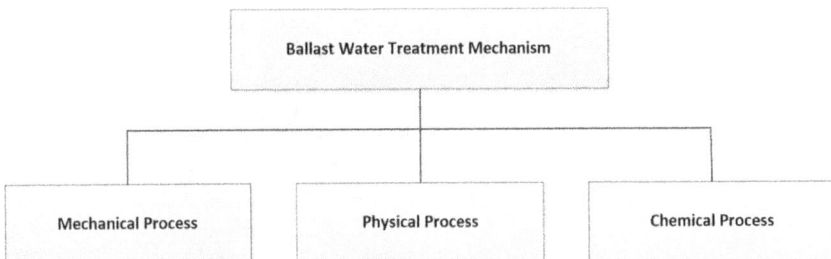

**FIGURE 8.2** The Ballast Water Treatment Mechanism.

Australia ports are the destination platform for some proposed procedures in dis-charging/charging the ballast water into this ocean.

## 8.2  LITERATURE REVIEW

### 8.2.1  BWM

The international convention of ballast water management plays an important role in controlling interchanging the wastewater from ships when travelling between the international ports. The IMO had adopted some stringent regulations for ships during the ballast water management process. The timeline of international ballast water management has been presented in Figure 8.3.

Some mandatory requirements of ballast water management have been issued and applied for all ships when they come to the American and European ports. On the other hand, the expanded evaluation of the BWM system has been conducted by Banerji et al., to prevent the dangers to humankind [16]. This proposed model has been provided the evaluative measurement for the ballast water discharged from ships through the usability and permissibility. Barbara et al. had proposed the emerging risk model based on the mathematical design and physical as well as chemical properties to describe the mandatory procedures of the water ballast discharging [17]. On the other hand, some research results have been published related to ballast water treatment methods [18]. It is a part of wastewater treatment methods from the indus-trial manufacture [19]. Moreover, the combined model has been proposed by Briski et al., to manage the ballast water discharging into the ocean sea environment [20]. The combination between the ballast water treatment and the ballast water exchange had some benefits to supervise the harmful species from the ballast water on ships.

### 8.2.2  THE BALLAST WATER TREATMENT TECHNOLOGIES

Before discharging the ballast water into the sea ocean environment, this water must be treated through some different methods. These methods could be listed as the mechanical treatment and the chemical treatment [21]. However, there is one tra-ditional method to charge/discharge the ballast water treatment on vessels through pumping in/out the wastewater from the international ports. The details of this pro-cess have been described in Figure 8.4.

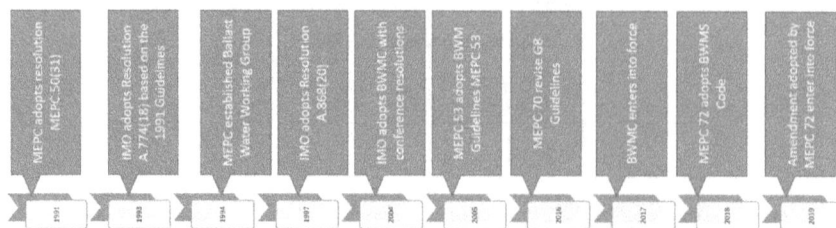

**FIGURE 8.3**  The Timeline of Events for Complement of Ballast Water Management by the International Maritime Organization.

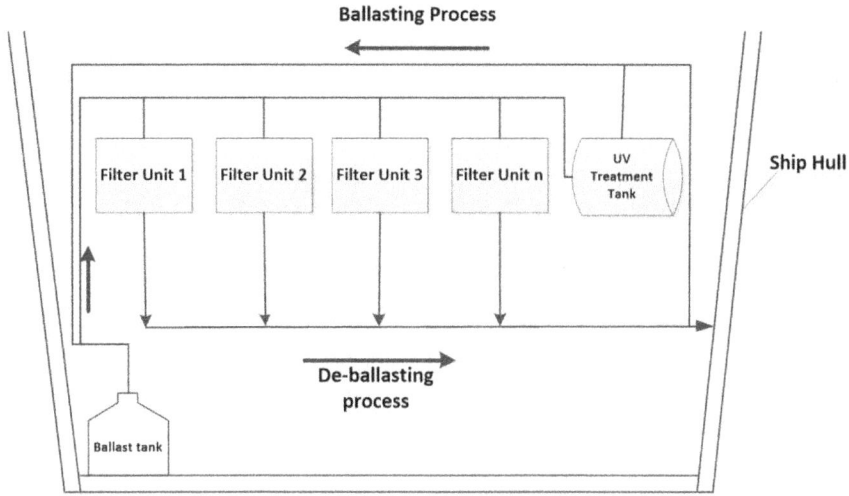

**FIGURE 8.4**  On-Board Ballast Water Treatment System.

There are some different techniques to treat the ballast water from ships. There-fore, the advanced technologies had been studied and combined between the chemical engineering and the environmental science so there are some research results which have been published in the international journals. The details of these technologies are described later.

### 8.2.3  RECENT WORKS RELATED TO THE BALLAST WATER TREATMENT

The appropriate selection of ballast water treatment system is a main research trend to help the stakeholders/ship owners to meet the stringent requirements of international maritime convention. There are some different approaches to select the appropriate ballast water system for the ship owner as well as saving the energy consumption of this system along with ensuring the reliable degree during the ship operation process. Tran had studied the ballast water treatment on a certain bulk carrier through proposing the actual treatment technique as well as the engines and equipment [21]. The hydrocyclonic treatment method had been studied in order to deal with the problems related to the ballast water treatment system [22]. Moreover, the advanced technology has been studied through combining between the PLC and LabVIEW [23]. Moreover, there are some conclusions from this research including (1) the filtration and the ultraviolet treatment method have been employed, (2) the hardware and software had been simulated for understanding the working operational principle of ballast water treatment system, and (3) the utilization of an actual ship must be conducted from the research results [23].

### 8.2.4  WASTEWATER TREATMENT TECHNOLOGY

There are a lot of research studies related to the ballast water management system through approaching both the mechanical engineering and the chemical engineering.

Especially, the ballast water management system is very important to effectively manage the wastewater system from the ocean-going vessels when they must be complied with strictly by the IMO. Consequently, the maritime transportation industry is multidisciplinary with a lot of requirements and international conventions that some countries in the world must respond to and meet through the maritime committee. The maritime transportation includes the onshore and offshore activities that they must meet and they must equip the sufficient devices as well as the novelty technologies to ensure effectively both the technology and the management.

Additionally, the ballast water management system will be divided into two kinds, including the offshore and onshore ballast system (Figure 8.5). This system is a specific scheme for treating the ballast water from ships. Normally, the onshore platform will include two main activities including the treatment facilities after de-ballasting and the treated-clean water during the ballast time period of ships. In contrast, the onboard ballast water system will also have two main activities including the onboard treatment and ballast water exchanging activities.

There are 596 vessels which have been tested and monitored by the Port State Control (PSC) about the technical condition during the operational process of ship [24]. There are some mandatory procedures to require the ship-owners and ship-operators (engine officers) to comply with some basic frameworks during this ballast water system. Moreover, it is mandatory that the ship-owners give some optimal policies to manage and

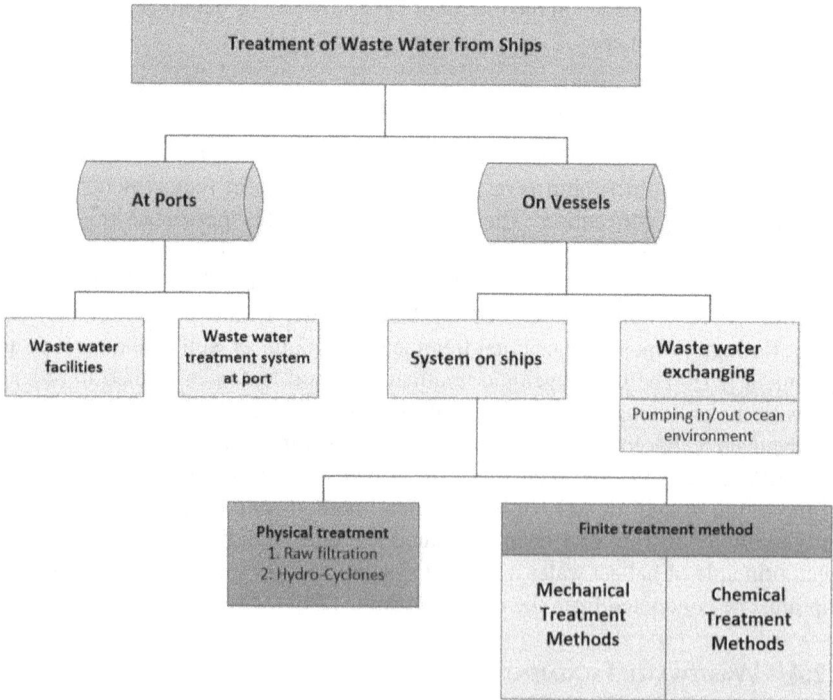

**FIGURE 8.5**   The Systematic Overview of BWM [21].

operate the ballast water system of ships. Therefore, there are some different studies which propose some methods to select the optimal ballast water management system.

There are two specific kinds of ballast water including the treatment system and the ballast/de-ballast water system. The popular treatment system is being used nowadays which is the ultraviolet (UV) system to treat the ballast water from ship (Figure 8.6).

## 8.3   CASE STUDY: WESTERN AUSTRALIA PORTS

The Western Australia ports have been selected to conduct this research. A lot of ocean-going vessels normally come to this port according to the charterer document of the ship owner. Then, the author had investigated and researched about this port when the ship comes to Australia. There are some policies related to this port which have adopted the "strategy manager" model in aim to ensure all facilities will be provided the best. The rate of investment is completed to meet the demand from the ship-owners of some countries in the world. On the other hand, the port authorities of the Western Australia have strong roles including the maritime logistics, the supply chain, and the risk management in port operation and management [26, 27]. The position of some Western Australia ports has been described in Figure 8.7. Normally, a number of ships

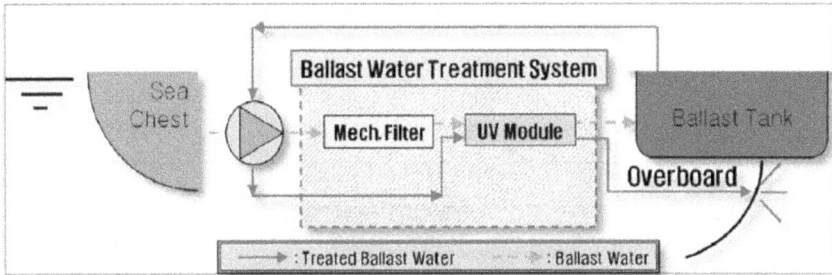

**FIGURE 8.6**   Ultraviolet (UV) Treatment Method on Ship [25].

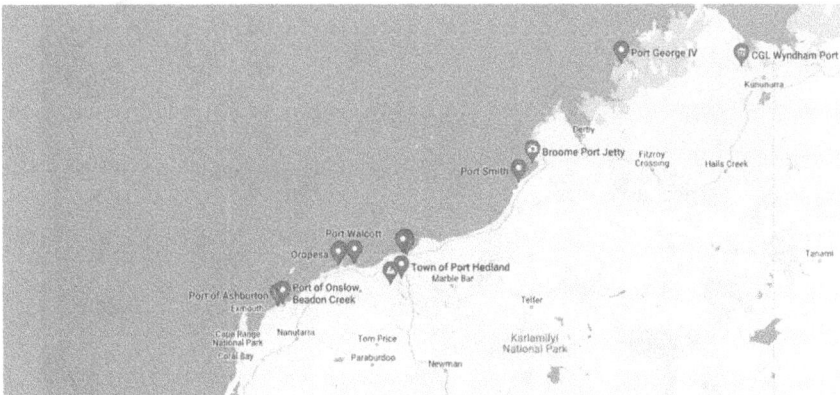

**FIGURE 8.7**   Some Western Australia Ports.

is at these ports such as Port Walcott, Port of Ashburton, Town of Port Hedland, Port Smith, etc. In Australia, there are a lot of ocean-going vessels coming to these ports and conducting the ballast/de-ballast water treatment from the different ocean areas. Therefore, this research has conducted the investigation of managing the ballast water treatment through a certain bulk carrier. The operational process had been studied and evaluated by the proposed theoretical methodologies from the existing research results.

The details of a certain bulk carrier have been provided in Table 8.1 and Figure 8.8. The process of ballast/de-ballast of this system had been managed by the fourth/third engine officer on vessels. This system will be supervised by the control panel like Figure 8.11. The operational process must be ensured and followed strictly

**TABLE 8.1**
**The Basic Parameter of a Certain Ship [21]**

| Number | Category | Coefficient |
|---|---|---|
| 1 | Name | LOCH MELFORT |
| 2 | IMO N$^O$ | 9658795 |
| 3 | Kind of ship | Bulk Carrier |
| 4 | L (m) | 176.83 |
| 5 | B (m) | 28.8 m |
| 6 | D (m) | 14.2 m |
| 7 | Draft | 9.826 m |
| 8 | DWT | 33,379 |

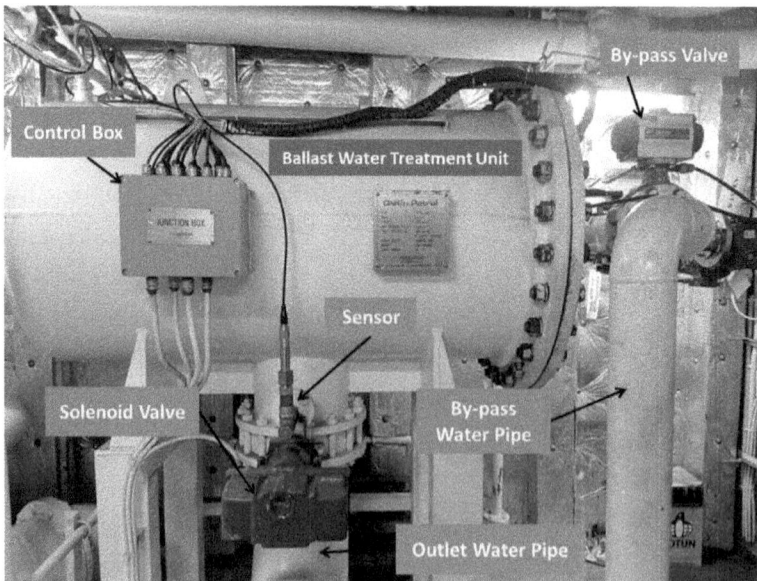

**FIGURE 8.8**   The Actual Treatment System on M/V LOCH MELFORT.

**FIGURE 8.9** Solenoid Valve.

**FIGURE 8.10** The Suction/Discharge Valve System.

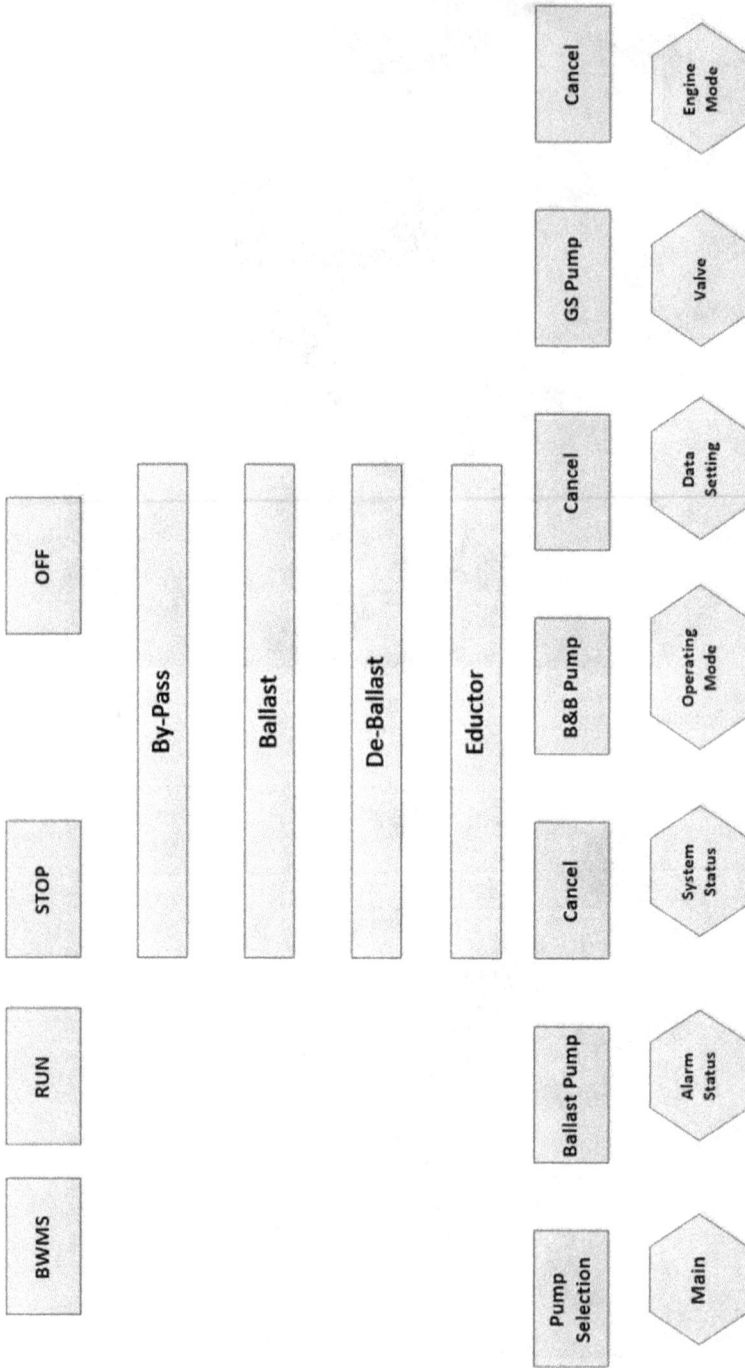

**FIGURE 8.11**    Controlling Panel on M/V LOCH MELFORT [21].

**TABLE 8.2**
**The Basic Parameters of System [21].**

| Category | Parameter |
|---|---|
| Model | GloEn-Patrol PF 750 |
| Capacity $(Q)$ $(Unit: m^3/h)$ | 750 |
| Maximum pressure $(P_{max})$ $(Unit: kgf/cm^2)$ | 7 |
| Pumping rate | 800 liters at $(P = 2$ $kgf/cm^2$ and $Q = 750$ $m^3/h)$ |
| Electrical source | 440 V (Alternative current) ~ Frequency: 60 Hz |

before discharging the ballast water into the ocean environment. The parameters of a ballast system have been provided in Table 8.2.

There are some mandatory steps that the ballasting/de-ballasting process have been conducted on a certain port of Western Australia. Therefore, this process has been carried out by the ballast pumps and the general pump. Normally, the kind of this pump is the centrifugal pump driven by the electrical motor. Therefore, the system will respond the mandatory requirements to ensure the water quality before discharging into the ocean environment. Moreover, the devices must be monitored and surveyed strictly before carrying out some procedures to pump into/out the ships. There are some further studied which would be conducted from this research result in the field of managing the ballast water from ships.

## 8.4 CONCLUDING REMARKS

The ballast water management system plays an important role in the field of waste-water treatment from the maritime transportation activities. There are a lot of different studies related to the ballast water treatment system through approaching the different aspects. Therefore, they are momentum to conduct this research through the systematic investigation from the recent ballast water treatment technology to utilize a certain bulk carrier in the maritime transportation. There are some advantages from this research to help the research scholars/stakeholders as a scenario to manage effectively the wastewater treatment system in aim with meeting the stringent requirements from the Ballast Water Management (BWM) IMO.

## ACKNOWLEDGMENTS

The author thanks the valuable comments from reviewers and editors to improve the quality of this research. Additionally, the author highly appreciates the chief engineer and engine officers of M/V LOCH MELFORT who helped me complete this research.

## REFERENCES

[1] IMO, 2007, Guidelines for risk assessment under regulation A-4 of the BWM convention (G7) resolution MEPC 162 (56) C.F.R.
[2] Briski E, et al., 2015, Combining ballast water exchange and treatment to maximize prevention of species introductions to freshwater ecosystems, *Environmental Science & Technology* 49(16), pp. 9566–9573.

[3] Seebens H, 2016, Predicting the spread of marine species introduced by global shipping, *Proceedings of the National Academy* 113(20), pp. 5646–5651.

[4] David M and Gollasch S, 2019, Risk assessment for ballast water management-Learning from the Adriatic Sea case study, *Marine Pollution Bulletin* 147, pp. 36–46.

[5] Saebi M, et al., 2020, Higher order patterns of aquatic species spread through the global shipping network, *PLoS One* 15(7), p. e0220353.

[6] Wang Z, et al., 2021, Integrated biological risk and cost model analysis supports a Geopolitical shift in Ballast water management, *Environmental Science & Technology* 55(19), pp. 12791–12800.

[7] Vorkapić A, et al., 2016, Shipboard ballast water treatment systems on seagoing ships, *Transactions on Maritime Science* 5(1), pp. 19–28.

[8] Darling JA, et al., 2018, Ballast water exchange and invasion risk posed by intracoastal vessel traffic: An evaluation using high throughput sequencing, *Environmental Science & Technology* 52(17), pp. 9926–9936.

[9] Carney KJ, et al., 2017, Evaluating the combined effects of ballast water management and trade dynamics on transfers of marine organisms by ships, *PLoS One* 12(3), p. e0172468.

[10] Tran TA, 2017, A research on the energy efficiency operational indicator EEOI calculation tool on M/V NSU JUSTICE of VINIC transportation company, Vietnam, *Journal of Ocean Engineering and Science* 2(1), pp. 55–60.

[11] Tran TA, 2019, Investigate the energy efficiency operation model for bulk carriers based on Simulink/Matlab, *Journal of Ocean Engineering and Science* 4(3), pp. 211–226.

[12] Tran TA, 2020, Effect of ship loading on marine diesel engine fuel consumption for bulk carriers based on the fuzzy clustering method, *Ocean Engineering* 207, p. 107383.

[13] Tran TA, 2021, Effects of the uncertain factors impacting on the fuel oil consumption of sea ocean-going vessels based on the hybrid multi criteria decision making method, *Ocean Engineering* 239, p. 109885.

[14] Lv B, et al., 2020, Vessel transport of antibiotic resistance genes across oceans and its implications for ballast water management, *Chemosphere* 253, p. 126697.

[15] Olsen RO, et al., 2016, Ultraviolet radiation as a ballast water treatment strategy: Inactivation of phytoplankton measured with flow cytometry, *Marine Pollution Bulletin* 103(1–2), pp. 270–275.

[16] Banerji S, et al., 2012, Assessing the risk of ballast water treatment to human health, *Regulatory Toxicology and Pharmacology* 62(3), pp. 513–522.

[17] Barbara W, et al., 2014, Emerging risks from ballast water treatment: The run-up to the international ballast water management convention, *Chemosphere* 112, pp. 256–266.

[18] David M and Gollasch S, 2015, *Global maritime transport and ballast water management: Issues and solutions.* Springer Book. ISBN: 978-94-017-9367-4, https://doi.org/10.1007/978-94-017-9367-4

[19] Roy S, et al., 2022, *Advanced industrial wastewater treatment and reclamation of water.* Springer Book. ISBN: 978-3-030-83811-9, https://doi.org/10.1007/978-3-030-83811-9

[20] Briski E, et al., 2015, Combining ballast water exchange and treatment to maximize prevention of species introductions to freshwater ecosystems, *Environmental Science & Technology* 49, pp. 9566–9573.

[21] Tran TA, 2022, Ballast water system treatment techniques in marine transportation industry: A case study of M/V LOCH MELFORT. In: Roy S, Garg A, Garg S and Tran TA (eds), *Advanced industrial wastewater treatment and reclamation of water.* Environmental science and engineering. Springer, Cham. https://doi.org/10.1007/978-3-030-83811-9_8

[22] Abu-Khader MM, et al., 2011, Ballast water treatment technologies: Hydrocyclonic a viable option, *Clean Technologies and Environmental Policy* 13, pp. 403–413.

[23] Wen-xiang C, et al., 2021, Research and development of ship ballast water treatment system, *IOP Conference Series: Earth and Environmental Science* 825, p. 012010.

[24] Pereira NN and Brinati HL, 2012, Onshore ballast water treatment: A viable option for major ports, *Marine Pollution Bulletin* 64(11), pp. 2296–2304.

[25] Jee J and Lee S, 2017, Comparative feasibility study on retrofitting ballast water treatment system for a bulk carrier, *Marine Pollution Bulletin* 119(2), pp. 17–22.

[26] Charlton HE, 1995, *The role of ports in Western Australia.* Principles to Guide Western Australia's Port Authority Development Through the 90's: Statement by the Minster for Transport, Perth, Australia.

[27] Özdemir Ü, 2022, A quantitative approach to the development of ballast water treatment systems in ships, *Ships and Offshore Structurers.* https://doi.org/10.1080/174453 02.2022.2077544

# 9 Phytoremediation of Hg and Cd Particulates Using Scenedesmus and Nostoc on Industrial Wastes

*Dr. M. Mathiyazhagan, Dr. G.Bupesh,*
*Dr. P. Visvanathan and P. Sudharsan*

## 9.1 INTRODUCTION

Water bodies are very much essential for all livelihoods and all kinds of industries. In recent years environmental pollutions and degradation of bioresourses are getting great attention at a global level due to global warming. However, water pollution, especially in lakes and rivers, is due to untreated industrial wastes. Industrial wastes contain biotoxic substances of heavy metals like Hg and Cd. These heavy metals are not only polluting the environment but also affect the livelihood of aquatic animals and humans. Metal toxicity in the natural environment becomes more important because of its non-biodegradability. Hence there is a need for a novel, efficient, eco-friendly, and cost effective method to remove the heavy metals from the water bodies.

Phytoremediation using algal strains like scenedesmus and nostoc could be a prospective alternative to conventional methods. Phytoremediation is the process where plants clean up contamination from soils, sediments, and water. Many reports reveal that the application of microbes for heavy metal treatment reduces the harmful effect of heavy metals [1–3]. Very recently nano green technology using microorganisms has become popular to reduce metallic toxic compounds from the environment [4]. However, phytoremediation has limitations because it greatly influences several factors such as temperature, pH, redox potential, moisture, nutrients, and composition of heavy metals [5]. So, single use of microbes for heavy metal removal is not effective due to competitiveness and excessive concentration of metals. Hence it can be improved by using combinations of techniques like bioabsorption, biosurfactants, bulking agents, and compost as well as biochar [6]. Therefore this study reviews a detailed investigation of previous works.

### 9.1.1 Macrophytes

Macrophytes are group of macroscopic plants present in or near to the water bodies in the form of submerged or floating condition. Macrophyte includes aquatic

DOI: 10.1201/9781003342830-9

flowering plants, ferns, bryophytes, and cyanobacteria. Aquatic macrophytes play a diverse role in the environment in the following ways;

- Being pollution bioindicators, macrophytes efficiently utilized minerals from the sediment nutrient pool.
- Submerged macrophytes of aquatic environments are used as green manure.
- Macrophytes form the basic energy source of food to the aquatic food chain.
- Macrophytes provide a suitable environment for breeding, nesting, and shelter for the aquatic animals including fish.

## 9.1.2 SCENEDESMUS

The algal species are good candidates for removing the pollutants from different sources. Due to their precise sensitivity over such contaminants they are used for detoxifying, monitoring, and as bioreactors with other algal or microbial organisms. Thus, they are useful in monitoring and evaluating different contaminated areas. Based on the production of intracellular dehydrogenase and urease enzymes on their specific growth condition, the algae could be used as bioindicators. *Scenedesmus* sp. is one of most useful species for bio reclamation processes. Research studies showed that *Scenedesmus* sp. is able to grow and remove Pb even in a lower concentration [7]. They also found that the lipids were accumulated in the algal cells and correlated them with their Pb removing efficiency (31% up to 1 mg/L concentration). Due to lipid production, those species could be also considered for the biofuel production. Research studies showed that *Scenedesmus* sp. anaerobically digested the palm oil mill effluent (POME) and effectively reduced the contaminated pollutants [8]. It reduced the ammonia, ammonium, phosphorus, and phosphate ions concentration up to 99.5%, 91.5%, 98.8%, and 97.2% respectively. Besides that it also reduced COD and BOD up to 50.5% and 61.6% respectively. Poly hydroxyalkanoates (PHAs) are serving as a natural carbon source that are present in many of the photosynthetic organisms and used as the nutrient source during the hostile situations. Due to their biochemical nature, they are easily biodegradable and have biocompatibility properties. Hence, the PHAs are used as the bio-based thermoplastics in everyday usages and constitute promising alternatives for the synthetic polymers. The microalgae are useful in synthesizing the PHAs in larger quantity and play a valuable role in waste management. Research studies showed that the production of PHAs was varying with the N, P, Fe, and salinity concentration in *Scenedesmus* sp. Their study showed that *Scenedesmus* sp. was a promising species to produce bio-resources in commercial systems [9].

## 9.1.3 NOSTOC

The *Nostoc* sp. has shown to oxidize and neutralize the different contaminants in wastewaters from various sources. Several studies showed that they are important candidates in processing of secondary wastewater treatment of industrial, municipal, and aquaculture origin [10]. Their ability to adopt to environmental variation such as UV radiation, higher sunlight intensity, desiccating niches, oxidative stress, and extreme temperature elevation help them to be more concurrent with the eco-nomical, ecofriendly candidates for bioremediation than other cyanobacterial spe-cies [11]. They ubiquitously grow on different culture conditions and have higher affinity towards the metals or xenobiotic compounds. They vary with their second-ary metabolite diversity characterized by the growing conditions such as duration of culture, exposure to the sun, attitude, and altitude of the sea water. Thus they have different chemicals with a wide variety of functional groups including hydroxyl, carboxyl, sulfite derivatives, and a number of charged groups that effectively bind with the contaminates for their neutralization [12–14]. The *Nostoc* sp. change their transcriptome during their blooming rather than the normal conditions. Their study revealed that elevation of genes for P-scavenging pathways strongly correlated with their phosphorus scavenging activity from the environment. In addition, they also showed that $N_2$ fixation was enhanced using nitrogen and phosphorus starvation. The cadminium (Cd) is the most noxious chemical that results from the industri-alization and is quite hard for environmental removal. As a contaminant, it causes severe defects on neurons and nutrient deficiency in the plants. Though several meth-ods such as ion exchange, chemical precipitation, electrocoagulation, electrodeposi-tion, and membrane filtration are used for the removal, it remains an unwanted guest in the water bodies. The cyanobacteria are very much useful in their removal and *N. muscorum* showed effective removal efficiency on the Cd ions. The study showed that *N. muscorum* is able to remove Cd, with 93.4% efficiency (0.5 to 0.033 mg L-1) in 21 days. Research study reported the decolorizing ability of the *Nostoc* sp. on methyl rage based on the hourly incubation [15]. They showed that the algal species achieved more 17.5% decolorizing efficiency within 5 hours and the maximum spe-cific decolorization rate was found to be 2.1734 mg $g^{-1}$ $h^{-1}$. Their study showed that the *Nostoc* sp. was a better alternative for decolorizing the azo based dyes. It was reported that the *Nostoc muscorum* showed effective degradation of textile effluents among other *cyanobacterial* species, *Anabaena variabilis, Oscillatoria salina,* and *Lyngbya majuscule* [16]. Several studies reported that *Nostoc* sp. produced high level of soluble cytoplasmic reductases known as azoreductase for reducing the azo dyes. Those enzymes convert the azo into simpler aromatic amines and lead to their deg-radation and decoloration of the effluents. Williams and Youngtor (2017) reported that the *Nostoc* and *Anabaena* combination worked well on the phytoremediation of the pesticide polluted river water. They studied the degradation of pesticide polluted water using different combinations of both cyanobacteria and the concentration of the contaminated water [17]. They found that AN3 combination (80% mixed culture) of *Anabaena* and *Nostoc* achieved approximately and their removal efficiency was in range between 95.9 and 99.9% for 40% and 91.7 and 100% for 80% respectively. These studies showed that the combination of *Nostoc* sp. with others also works well for phytoremediation processes [18, 19].

## 9.2   TOXICITY OF HEAVY METALS

Various industries released various heavy metals into the natural water bodies. Any kind of heavy metals released into the environment is carcinogenic or toxic in nature. So removal of heavy metals from wastewater is a tedious problem. In this regard different

**TABLE 9.1**
**Common Heavy Metals Present in the Environment**

| S.No | Common Heavy Metal Pollutants |
|------|-------------------------------|
| 1. | Cadmium (Cd) |
| 2. | Chromium Cr) |
| 3. | Copper (Cu) |
| 4. | Mercury (Hg) |
| 5. | Lead (Pb) |
| 6. | Zinc (Zn) |
| 7. | Nickel (Ni) |
| 8. | Cobalt (Co) |
| 9. | Magnesium (Mg) |

types of adsorption process which are cost effective and ecofriendly are widely used for removal of heavy metals. Different methods are utilized to decontaminate the environment from these kinds of contaminants. But most are expensive and not as per their optimum performance. The chemical technologies create a great amount volumetric sludge and increase the prices; thermochemical methods are both tedious and costly and all of these methods can also destroy the main components of soil.

Generally, decontamination of heavy metals from the contaminated soils involves a method of disposal with the hazards associated with transportation of polluted soil and removal of contaminants from a landfill into an adjacent environment.

Heavy metals are elements with metallic properties and an atomic number greater than 20. The common heavy metal pollutants are present in the environment. Heavy metals are micronutrients essential for plant growth, tabulated in Table 9.1.The pollutant heavy metals such as Pb, Co, and Cd can be separated from other pollutants, as they cannot be degraded biologically but can be contaminated in micro and macro organisms, thus affecting various diseases and disorders even in relatively lower concentrations. The toxic heavy metals have an effect on plant growth and have a negative effect on soil organisms. It is well known that heavy metals cannot be chemically degraded and need to be remediated physically or be converted into nontoxic compounds.

## 9.2.1 Mercury

Mercury is a naturally occurring metal present in different forms. According to physical nature, mercury is a shiny, silver-white, odorless liquid. Mercury chemically reacts with other nonmetals such as chlorine, sulfur, or oxygen, to produce different compounds or salts, which are mostly available in the form of powders or crystals. Mercury also reacts with carbon to form organic mercury compounds. As any other metal, mercury is present in the soil in different forms and it dissolves as a free ion or soluble complex. Sometimes it is adsorbed by combining because of electrostatic forces, chelation, and precipitation such as sulphide, carbonate, hydroxide, and phosphate.

Among three soluble forms of Hg, the most reduced form is $Hg^0$ metal with the other two forms being ionic of mercurous ion and mercuric ion $Hg^{2+}$ in oxidizing conditions (low pH). As $Hg^+$ ion is unstable under environmental conditions it dissociates into $Hg^0$ and $Hg^{2+}$. The greatest potential route for the conversion of mercury in the soil is methylation to methyl or dimethyl mercury using anaerobic bacteria.

Mercury is a prominent contaminant having bioaccumulation ability in fish, animals, and human beings. Mercury salts and organo mercury compounds are highly toxic substances in our surroundings. The chemical reaction mechanism and pollution index mainly depend on the oxidation state of mercury in the compound [20–22].

The mercury contamination is mostly affected by various industries such as petrochemicals, painting, and conventional uses of agricultural pesticides and fungicides. The most common other sources of mercury available in our surroundings are tabulated in Table 9.2

Mercury ions produce contamination effects by precipitating with protein, inhibiting enzymes, and mostly corroding other substances. Mercury not only combines with sulfhydryl groups but also with phosphoryl, carboxyl, amide, and amine groups. Proteins are very vulnerable to reaction with mercury. The predominant pathway of exposure to methylmercury [23–26] is through the ingestion of fish. After absorption of methylmercury, it penetrates into erythrocytes where it will affect hemoglobin in the blood more than 90%. In the human brain, methylmercury causes focal necrosis of neurons and destruction of glial cells and is highly pollutant to the cerebral and cerebellar cortex.

Mercuric chloride containing waste had been discharged into the bays and it became contaminated in the fish after chemical conversion to methylmercury by plankton [27, 28].

Dimethylmercury is the highly toxic form of mercury that is dangerous after accidental exposure. Dimethylmercury [29] is a volatile liquid organic mercuric compound, used as a reference material in nuclear magnetic resonance chemistry

**TABLE 9.2**
**Sources of Mercury Present in the Surroundings**

| S.No | Common Mercury Sources |
| --- | --- |
| 1. | Household bleach |
| 2. | Battery acid |
| 3. | Caustic chemicals |
| 4. | Household lye |
| 5. | Manometers |
| 6. | Thermometers |
| 7. | Barometers |
| 8. | Dental amalgam |
| 9. | Pesticides |
| 10. | Medical instruments |
| 11. | Pharmaceuticals products |

laboratories. Before death, dimethylmercury can cause neurological damage such as loss of audio logical (speech recognition) systems. Research studies revealed that dimethyl mercury [30] must be metabolized to methylmercury prior to entering the brain [31, 32].

### 9.2.2 CADMIUM

Cadmium is a potential toxic heavy metal with destructive impact on human organ systems. Detoxification of cadmium with EDTA and other chelators is therapeutically beneficial in humans and animals using established protocols.

Human exposure to Cd occurs chiefly through inhalation or ingestion depending on particle size and absorption through skin contact is negligible. It is used for intestinal absorption which is greater in persons with iron, calcium, or zinc deficiency [33, 34]. Specially, cigarette smoking is the most significant source of human cadmium exposure [35]. Blood and kidney Cd levels are consistently higher in smokers compared to nonsmokers. Industrial exposure can be significant in occupational settings, as an example, welding or soldering, and it can produce severe chemical pneumonitis [36].

Cadmium exposure occurs from ingestion of contaminated food or water and can produce long-term health effects. Drugs and dietary supplements may also be a source of contamination [37]. Cadmium induces tissue injury through creating oxidative stress [38–40], epigenetic changes in DNA expression [41–43], inhibition, or up regulation of transport pathways [44–46] particularly in the proximal S1 segment of the kidney tubule. Different pathologic mechanisms include competitive interference with the physiologic action of Zn or Mg [47–49], inhibition of heme synthesis [50], and impairment of mitochondrial function, potentially inducing apoptosis [51]. Also depletion of glutathione (structural distortion of proteins) has been observed due to Cd binding to sulfhydryl groups [52]. 30% of body cadmium is deposited in the kidney tubule region, with tubular damage being proportionate to the quantity of cadmium not bound to metallothionein. Diabetics are more vulnerable to renal tubular damage from Cd exposure [53]. Cadmium is responsible for impairment of Vitamin D metabolism in the kidney [54], with deleterious impact on bone, which results in osteomalacia and/or osteoporosis. The best example of this process is *itai-itai* disease in Japan, which combines severe pain from osteomalacia with osteoporosis, renal tubular dysfunction, anemia, and calcium malabsorption [55].

### 9.3 CONCLUSION

Phytoremediation is a low cost effective technology and an alternative to engineered technologies. The *Scenedesmus*- and *Nostoc*- based eco restoration attract huge research interest due to ease maintenance and rapid re-growing after the first treatment. Further, this type of low-cost strategy can be easily adopted by the people through Phyto-based approaches, since they usually create greenish and more aesthetic views than the artificial technologies and remain a source of valuable by-products such as biofuel etc. However, the study demonstrated that the use of algae can be a good solution for remediation of heavy metals. Further there is need of physio chemical cum kinetics of the growth parameters and associated characters for effective restoration.

## REFERENCES

[1]    R. J. Ndeddy Aka, and O. O. Babalola (2016) "Effect of bacterial inoculation of strains of pseudomonas aeruginosa, alcaligenes feacalis and bacillus subtilis on germination, growth and heavy metal (Cd, Cr, and Ni) uptake of Brassica juncea," *International Journal of Phytoremediation*, vol. 18, no. 2, pp. 200–209.

[2]    H. S. Abbas, M. I. Ismail, M. T. Mostafa, et al. (2014) "Biosorption of heavy metals: A review," *Journal of Chemical Science and Technology*, vol. 3, pp. 74–102.

[3]    A. Akcil, C. Erust, S. Ozdemiroglu, et al. (2015) "A review of approaches and techniques used in aquatic contaminated sediments: Metal removal and stabilization by chemical and biotechnological processes," *Journal of Cleaner Production*, vol. 86, pp. 24–26.

[4]    T. Klaus-Joerger, R. Joerger, E. Olsson, et al. (2001) "Bacteria as workers in the living factory: Metal-accumulating bacteria and their potential for materials science," *Trends in Biotechnology*, vol. 19, no. 1, pp. 15–20.

[5]    P. K. Shukla, S. Sharma, K. N. Singh, et al. (2013) *Rhizoremediation: A Promising Rhizosphere Technology*, Y. B. Patil, and P. Rao, Eds., vol. 331. IntechOpen, London, United Kingdom.

[6]    A. Wiszniewska, E. Hanus-Fajerska, E. MuszyŃska, et al. (2016) "Natural organic amendments for improved phytoremediation of polluted soils: A review of recent progress," *Pedosphere*, vol. 26, no. 1, pp. 1–12.

[7]    T.-L. Pham, T.-S. Dao, H. N. Bui, et al. (2020) "Lipid production combined with removal and bioaccumulation of Pb by Scenedesmus sp. green alga," *Polish Journal of Environmental Studies*, vol. 29, no. 2.

[8]    K. F. Kamarudin, Z. Yaakob, R. Rajkumar, et al. (2013) "Bioremediation of palm oil mill effluents (POME) using Scenedesmus dimorphus and Chlorella vulgaris," *Advanced Science Letters*, vol. 19, no. 10, pp. 2914–2918.

[9]    G. García, J. E. Sosa-Hernández, L. I. Rodas-Zuluaga, et al. (2021) "Accumulation of PHA in the microalgae Scenedesmus sp. under nutrient-deficient conditions," *Polymers*, vol. 13, no. 1, p. 131.

[10]   W. N. A. Kadir, M. K. Lam, Y. Uemura, et al. (2018) "Harvesting and pre-treatment of microalgae cultivated in wastewater for biodiesel production: A review," *Energy Conversion and Management*, vol. 171, pp. 1416–1429.

[11]   V. A. Mantri, M. A. Kazi, N. B. Balar, et al. (2020) "Concise review of green algal genus Ulva Linnaeus," *Journal of Applied Phycology*, vol. 32, no. 5, pp. 2725–2741.

[12]   S. Verma, and A. Kuila (2019). "Bioremediation of heavy metals by microbial process," *Environmental Technology & Innovation*, vol. 14, p. 100369.

[13]   D. Kour, T. Kaur, R. Devi, et al. (2021) "Beneficial microbiomes for bioremediation of diverse contaminated environments for environmental sustainability: Present status and future challenges," *Environmental Science and Pollution Research*, pp. 1–23.

[14]   J. Lu, B. Zhu, I. Struewing, et al. (2019). "Nitrogen–phosphorus-associated metabolic activities during the development of a cyanobacterial bloom revealed by metatranscriptomics," *Scientific Reports*, vol. 9, no. 1, pp. 1–11.

[15]   M. H. Hussein, G. S. Abou El-Wafa, S. A. Shaaban-Dessuki, et al. (2018). "Bioremediation of methyl orange onto Nostoc carneum biomass by adsorption, kinetics and isotherm studies," *Global Advanced Research Journal of Microbiology*, vol. 7, no. 1, pp. 6–22.

[16]   S. D. Noel, and M. R. Rajan (2014) "Cyanobacteria as a potential source of Phycoremediation from textile industry effluent," *Journal of Bioremediation & Biodegradation*, vol. 5, p. 260.

[17] T. Robinson, G. McMullan, R. Marchant, et al. (2001) "Remediation of dyes in textile effluent: A critical review on current treatment technologies with a proposed alternative," *Bioresource Technology*, vol. 77, no. 3, pp. 247–255.

[18] I. M. C. Gonclaves, A. Gomes, R. Bras, et al. (2000) "Biological treatment of effluent containing textile dyes," *Colouration Technology*, vol. 116, pp. 393–397.

[19] A. El-Hameed, M. Mohamed, M. Abuarab, et al. (2021) "Phytoremediation of contaminated water by cadmium (Cd) using two cyanobacteria species (Anabaena Variabilis and Nostoc Muscorum)," *Environmental Sciences Europe*, vol. 33, p. 135.

[20] N. Takahata, H. Hayashi, S. Watanabe, et al. (1970) "Accumulation of mercury in the brains of two autopsy cases with chronic inorganic mercury poisoning," *Folia Psychiatrica et Neurologica Japonica*, vol. 24, pp. 59–69.

[21] H. A. L. Rothstein (1960) "The metabolism of mercury in the rat studied by isotope techniques," *Journal of Pharmacology and Experimental Therapeutics*, vol. 130, pp. 166–176.

[22] N. Matsuo, T. Suzuki, and H. Akagi (1989) "Mercury concentration in organs of contemporary Japanese," *Archives of Environmental Health*, vol. 44, pp. 298–303.

[23] L. E. Karper, N. Ballatori, and T. W. Clarkson (1992) "Methylmercury transport across the blood-brain barrier by an amino acid carrier," *American Journal of Physiology*, vol. 267, pp. R761–R765.

[24] T. G. Kershaw, T. W. Clarkson, and P. H. Dhahir (1980) "The relationship between blood-brain levels and dose of methylmercury in man," *Archives of Environmental Health*, vol 35, pp. 28–36.

[25] M. Harada (1995) "Minamata disease: Methylmercury poisoning in Japan caused by environmental pollution," *Critical Reviews in Toxicology*, vol. 25, pp. 1–24.

[26] F. Bakir, S. F. Damluji, L. Amin-Zaki, et al. (1973) "Methyl mercury poisoning in Iraq," *Science*, vol. 181, pp. 230–241.

[27] D. W. Nierenberg, R. E. Nordgren, M. B. Chang, et al. (1998) "Delayed cerebellar disease and death after accidental exposure to dimethylmercury," *The New England Journal of Medicine*, vol. 338, pp. 1672–1676.

[28] F. E. Musiek, and D. P. Hanlon (1999) "Neuroaudiological effects in a case of fatal dimethylmercury poisoning," *Ear & Hearing*, vol. 20, pp. 271–275.

[29] T. G. Kershaw, T. W. Clarkson, and P. H. Dhahir (1980) "The relationship between blood-brain levels and dose of methylmercury in man," *Archives of Environmental Health*, vol. 35, pp. 28–36.

[30] D. W. Nierenberg, R. E. Nordgren, M. B. Chang, et al. (1998) "Delayed cerebellar disease and death after accidental exposure to dimethylmercury," *The New England Journal of Medicine*, vol. 338, pp. 1672–1676.

[31] F. E. Musiek, and D. P. Hanlon (1999) "Neuroaudiological effects in a case of fatal dimethylmercury poisoning," *Ear & Hearing*, vol. 20, pp. 271–275.

[32] K. Ostlund (1969) "Studies on the metabolism of methylmercury in mice," Acta *Pharmacology & Toxicology*, vol 27, pp. S1–S132.

[33] U. S. Geological Survey (2012) *Mineral Commodity Summaries*, U.S. Geological Survey, Rolla, Mo, USA.

[34] G. F. Nordberg, K. Nogawa, M. Nordberg, et al. (2007) "Cadmium," in *Chapter 23 in Handbook of the Toxicology of Metals*, G. F. Nordberg, B. F. Fowler, M. Nordberg, and L. Friberg, Eds., 3rd edition, pp. 445–486, Elsevier, Amsterdam, The Netherlands.

[35] L. Friberg (1983) "Cadmium," *Annual Review of Public Health*, vol. 4, pp. 367–373.

[36] D. R. Abernethy, A. J. DeStefano, T. L. Cecil, et al. (2010) "Metal impurities in food and drugs," *Pharmaceutical Research*, vol. 27, no. 5, pp. 750–755.

[37] V. Matović, A. Buha, Z. Bulat, et al. (2011) "Cadmium toxicity revisited: Focus on oxidative stress induction and interactions with zinc and magnesium," *Arhiv za Higijenu Rada i Toksikologiju*, vol. 62, no. 1, pp. 65–76.

[38] R. C. Patra, A. K. Rautray, and D. Swarup (2011) "Oxidative stress in lead and cadmium toxicity and its amelioration," *Veterinary Medicine International*, vol. 2011, Article ID 457327.

[39] A. Cuypers, M. Plusquin, T. Remans et al. (2010) "Cadmium stress: An oxidative challenge," *BioMetals*, vol. 23, no. 5, pp. 927–940.

[40] B. Wang, C. Shao, Y. Li, et al. (2012) "Cadmium and its epigenetic effects," *Current Medicinal Chemistry*, vol. 19, no. 16, pp. 2611–2620.

[41] R. Martinez-Zamudio, and H. C. Ha (2011) "Environmental epigenetics in metal exposure," *Epigenetics*, vol. 6, no. 7, pp. 820–827.

[42] C. Luparello, R. Sirchia, and A. Longo (2011) "Cadmium as a transcriptional modulator in human cells," *Critical Reviews in Toxicology*, vol. 41, no. 1, pp. 75–82.

[43] F. Thévenod (2010). "Catch me if you can! Novel aspects of cadmium transport in mammalian cells," *BioMetals*, vol. 23, no. 5, pp. 857–875.

[44] L. Wan, and H. Zhang (2012) "Cadmium toxicity: Effects on cytoskeleton, vesicular trafficking and cell wall reconstruction," *Plant Signaling & Behavior*, vol. 7, no. 3, pp. 345–348.

[45] E. Van Kerkhove, V. Pennemans, and Q. Swennen (2010) "Cadmium and transport of ions and substances across cell membranes and epithelia," *BioMetals*, vol. 23, no. 5, pp. 823–855.

[46] D. A. Vesey (2010) "Transport pathways for cadmium in the intestine and kidney proximal tubule: Focus on the interaction with essential metals," *Toxicology Letters*, vol. 198, no. 1, pp. 13–19.

[47] M. Abdulla, and J. Chmielnicka (1989) "New aspects on the distribution and metabolism of essential trace elements after dietary exposure to toxic metals," *Biological Trace Element Research*, vol. 23, pp. 25–53.

[48] J. M. Moulis (2010) "Cellular mechanisms of cadmium toxicity related to the homeostasis of essential metals," *BioMetals*, vol. 23, no. 5, pp. 877–896.

[49] G. S. Shukla, and R. L. Singhal (1984). "The present status of biological effects of toxic metals in the environment: Lead, cadmium, and manganese," *Canadian Journal of Physiology and Pharmacology*, vol. 62, no. 8, pp. 1015–1031.

[50] A. Schauder, A. Avital, and Z. Malik (2010) "Regulation and gene expression of heme synthesis under heavy metal exposure—review," *Journal of Environmental Pathology, Toxicology and Oncology*, vol. 29, no. 2, pp. 137–158.

[51] G. Cannino, E. Ferruggia, C. Luparello, et al. (2009) "Cadmium and mitochondria," *Mitochondrion*, vol. 9, no. 6, pp. 377–384.

[52] M. Valko, H. Morris, and M. T. D. Cronin (2005) "Metals, toxicity and oxidative stress," *Current Medicinal Chemistry*, vol. 12, no. 10, pp. 1161–1208.

[53] A. Åkesson, T. Lundh, M. Vahter et al. (2005) "Tubular and glomerular kidney effects in Swedish women with low environmental cadmium exposure," *Environmental Health Perspectives*, vol. 113, no. 11, pp. 1627–1631.

[54] T. Kjellström (1992). "Mechanism and epidemiology of bone effects of cadmium," *IARC Scientific Publications*, no. 118, pp. 301–310.

[55] T. Ogawa, E. Kobayashi, Y. Okubo, et al. (2004) "Relationship among prevalence of patients with Itai-itai disease, prevalence of abnormal urinary findings, and cadmium concentrations in rice of individual hamlets in the Jinzu River basin, Toyama prefecture of Japan," *International Journal of Environmental Health Research*, vol. 14, no. 4, pp. 243–252.

# 10 Employment of Phytal Mediated Metallic Nanoparticles for the Effective Remediation of Wastewater Particulates

*P. Sudharsan, T. Siva Vijayakumar, G. Bupesh,*
*Bhagyudoy Gogoi and M. Mathiyazhagan*

## 10.1 INTRODUCTION

The top-down and bottom-up approaches to manufacturing iron nanoparticles are the two basic approaches. Physical methods are used to make nanoparticles in top-down techniques [1]. Bottom-up techniques necessitate high temperatures, pressures, or energy requirements; are hence relatively expensive; and have a number of draw-backs. Because there is a growing desire for eco-friendly, low-cost, and less harmful ways of synthesising nanoparticles, bio-synthesis is presented as a practicable sub-stitute to conventional techniques [2–4]. Microorganisms like bacteria, algae, fungus, yeast, and plant extracts will play a role in the production of nanoparticles from naturally biodegradable substances (Figure 10.1).

### 10.1.1 Nanoparticle Synthesis and Wastewater Treatment

Green nanoparticle (NP) synthesis is a new research topic in green nanotechnology. Compared to other standard physiochemical procedures, this technology is safe or less harmful, efficient for the environment, affordable, and effective [5]. Due to their excellent efficacy, selections of eco-friendly nanoparticles are being employed more often to treat wastewater and biocompatibility. For the green production of zinc oxide nanoparticles, DPW pulp was employed as an efficient bio-reductant (ZnO-NPs) [6]. For DPW-mediated ZnO/NPs synthesis, an easy and environmentally friendly approach with minimal reaction time and temperature was used. Zone-of-inhibition as evaluated by the antimicrobial technique, DP-ZnOs, showed substantial anti-bacterial activity on a variety of harmful microorganisms [7–9]. The photocatalytic destruction of harmful methylene blue and eosin dyes with these nanoparticles yielded a 90% degradation efficiency [10, 11]. With the crude extract of jatropha cur-cas and cinnamomum tamala herbal extract, magnetite nanoparticles were manufac-tured in an open-air setting in a cheaper and greener approach (CT). FE-SEM photos

DOI: 10.1201/9781003342830-10

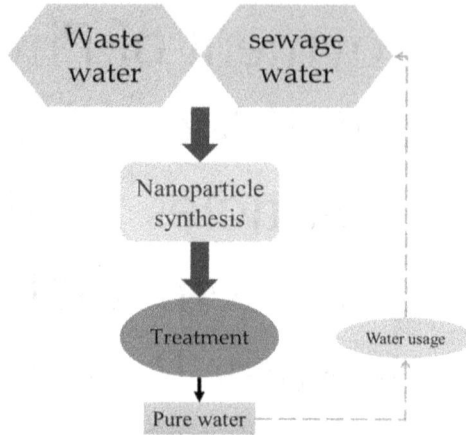

**FIGURE 10.1**   Wastewater Treatment by Nanoparticles.

revealed that the synthesised MNPs had size ranges of 20–42 nm for JC-Fe3O4 and 26–35 nm in SEM images [12].

Synthetic magnetic nanoparticles were created synthetically and their impact on treating wastewater, dye adsorption, potentially toxic removal, and bactericidal, oxidative, and cytotoxic activities were studied. The efficacy of calcined green produced Fe/Ni nanoparticles for As removal from aqueous solution was tested in this work. Under ideal experimental circumstances, C–Fe/Ni NPs had an efficiency of 87.3%. The removal is greatly hampered by the presence of coexisting anions, including Cl, $NO_3$, SO42, $H_2PO_4$, and $HCO_3$ (Figure 10.2).

This study looked at how green-synthesised nanocatalysts and nanomaterials can be used to clean wastewater and purify water in the long run. They have a huge area, mechanical properties, high chemical reactions, and low energy consumption, making them ideal for removing dangerous chemicals and pollutants from water (Figure 10.3). This chapter outlines the benefits and drawbacks of using these nanomaterials for industrial wastewater cleanup, as well as key challenges. Recent advances in nanofabrication, including intelligent and smart nanofabrication, highlight the field's promise for wastewater treatment and recovery. The nanotechnology industry promotes nano as a "green" technology that will help current companies enhance their environmental performance [13–15].

Three parts of a green nanoparticle production technique should be considered as solvent, capped agent, and reducing agents. The availability of polyphenols in biodegradable materials is primarily responsible for the decomposition of any organic materials by the green method. It is recommended that microcrystalline cellulose be subjected to a straightforward and ecologically friendly $Fe_2+/H_2O_2$ oxidation in order to produce cellulose nanocrystals with a high carboxylation concentration [16]. These CNCs can effectively adsorb up to 95.6% of dyes and 82.3% of metal ions from wastewater. Table 10.1 displays characteristics of nanoparticles.

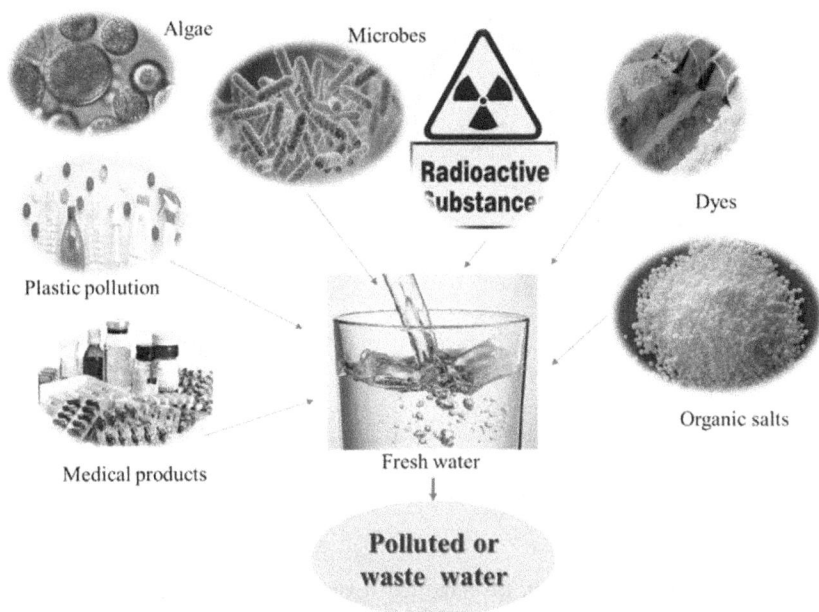

**FIGURE 10.2** Formation of Wastewater.

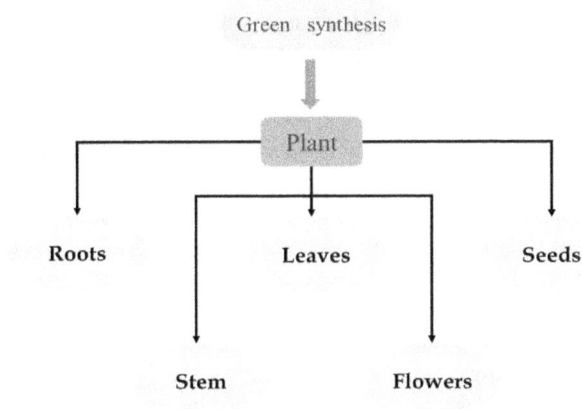

**FIGURE 10.3** Green Synthesis of Nanoparticle.

**TABLE 10.1**
**Characteristics of Nanoparticles**

**Green Nanoparticle Characteristics**

| | |
|---|---|
| Low cost | Non-toxic to living beings |
| Eco-friendly | Simple procedure |
| Easy methods | Easily available |

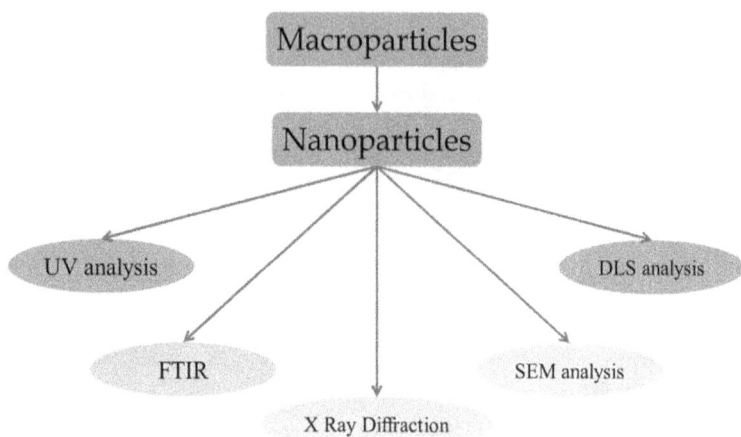

**FIGURE 10.4**    Nanoparticle Characterisation.

The goal of this research is to make leaf extracts, precursors, and iron nanos-tructures, as well as to test their usefulness in treating home wastewater. Phosphate removal rate was 98.08%, ammonia nitrogen removal rate was 84.32%, and chemical oxygen demand removal rate was 82.35% for azadiracta indica. When compared to other plant extracts for treating residential wastewater, the overall performance of this leaf extract was good. The green technique was adopted in this work to produce pure and doped copper oxide nanoparticles. Several characterisation techniques, such as EDX, XRD, UV/VIS, PL, SEM, and FTIR, are employed to investigate their physiological, biochemical, and optical properties [17, 18]. The produced compound is employed as a photocatalytic for methylene blue (MB) dye degradation. Doping lowers the rate of electron pair recombination during photocatalysis, making it a possible photocatalytic agent. The maximum degrading efficiency for 2% La doped CuO-NP is 84% in 150 minutes, while for 3% it is 75% in 75 minutes. The simulation is fairly close to the experimental findings, according to a comparative examination of rate constants. PAHs are organic pollutants produced mostly by the incomplete combination of organic substances, including gas, oil, and charcoal. PAHs may harm the environment and living creatures in irreversible ways. Several methods have been developed in recent decades with the goal of identifying and treating these chemicals in wastewater. This study focuses on adsorption processes using various adsorbent particles, with a particular focus on graphene nanomaterials manufactured using tra-ditional and green methods (Figure 10.4).

## 10.2   GREEN SYNTHESIS OF NANOPARTICLES

Green nanoparticle (NP) synthesis has evolved as an environmentally responsible method of producing nanomaterials with a wide range of properties. Bacteria, fungus, and algae are among the biological groups involved in the manufacture of NPs. Algal cells may ingest heavy metals and use them as a nutrition source to build biomass through physiological and biological mechanisms [19]. Nano-zerovalent iron per-formed brilliantly as an electron donor in the decrease and removal of nitrates. NZVI's simple aggregation and weak oxidation impeded nitrate removal [20]. It was created and characterised as a novel polyphenol and modified polypropylene carrier composite packing. The oxidised nano-zerovalent iron was transformed to Fe2+/Fe3+, which had a higher affinity for nitrate adsorption and co-precipitation. Adsorption techniques are widely utilised to eliminate specific contaminants in dyes, aniline, and treatment of wastewater procedures. Furfural is a contaminant that is organic in origin and has a hazardous impact on individuals and the environment, notably marine creatures. Pomegranate fruit peel extract was employed to make acti-vated carbon, as well as a carbon nanostructure coated with zerovalent iron. The nanoparticles have a surface area porosity structure of 821.74 m2/g.

Hydrothermal treatment of sewage sludge produced nanorods with diameters of 80 nm and lengths of a 100 nm. The pilot-scale conversion produced no secondary waste, and the synthesised products' adsorption effectiveness to wastewater with a high starting content of 1,000 m/L was above 99.5%. Hexavalent chromium replaced the single bond OH/single bond SH group of nearby structural iron to generate a steady ligand during wastewater treatment. A green technique was used to make core shell nanoparticles that are very efficient in dye degradation [21]. They degrade rhodamine B in the same way that photocatalysts do, with a degradation ratio of less than 96.0%. A green approach for fabricating hybrid structures might pave the way for the development of innovative photocatalysts with wastewater treatment applica-tions. *Wedelia urticifolia* leaf extract was utilised to make copper oxide nanoparticles (CuONPs), which were employed as adsorbents for the Rhodamine-b (RhB) dye [22]. Fourier transform infrared analysis and UV–Visible spectroscopy were used to fore-cast the biomolecules responsible for the synthesis [23]. CuONPs as-synthesised have a high capacity for RhB adsorption, making them possible alternatives to exist-ing wastewater treatment methods [24].

Onion peel was used to produce ZnONPs, which were then analysed utilising advanced equipment such as FTIR, DLS, and FESEM. The particles were subse-quently employed as plant nutrients, and their impact on the development of beans and wheat was assessed. The germination and seedling growth of those seeds were found to be superior to the control seed. When compared to monometallic NPs, sil-ver/palladium nanoparticles efficiently absorb UV light and have higher photocat-alytic processes [25]. For the bio-produced Ag and Ag/Pc NPs, Bacillariophyceae (diatom) algae extract was selected for this study and obtained using a statistical database. The particle shape and size were determined to be 27 and 56 nm, respec-tively, with a spherical shape, and the average crystalline size was calculated to be 23 nm and 56 nm for Ag NPs, respectively. In today's world, scientists are faced with the task of purifying water in an environmentally responsible and cost-effective manner. The current work describes a green production technique for ZnO-NPs using S. cumini plant leaf extract. The synthesised NPs degraded RhB dye by 98% and purified dye-polluted water by 98%, indicating that they might be used as a dye degradation catalyst. They were also used as a nano-catalyst for Rhodamine B (RhB) dye degradation and removal. A natural synthesis approach was successfully synthesised using $TiO_2$ NPs as a simple synthetic technique using three separate aqueous pomegranate leaf extracts, beta, and Seder. For contrast, typical $TiO_2$NPs are produced utilising a straightforward wet chemical method [26]. All of the samples that were created either biologically or chemically were examined using various methods for characterising materials such TEM, XRD, FT-IR, and UV-Vis spectroscopy. According to the results, the type of plant extract is predisposed to altered particle properties such photocatalysts, particle size, and phases ratio.

## 10.3   CONCLUSION

The phytal mediated synthesis of nanoparticles is increasingly being used in treating wastewater due to its high efficiency and biocompatibility. Green nanoparticles are highly recyclable and are being used to remove metal ions from wastewater without compromising their stability, as well as degrading a variety of organic pollutants in wastewater and being treated for reuse and recycling, potentially resolving various water quality issues around the world. The study demonstrated that treatment of waste water by using green synthesised metallic nanoparticles can be a promising strategy to remediate the elevated level of pollutants up to the permissible level in the environment.

## REFERENCES

[1]   Zhou, Y., & Li, X. (2022). Green synthesis of modified polyethylene packing supported tea polyphenols-NZVI for nitrate removal from wastewater: Characterization and mechanisms. *Science of the Total Environment*, 806, 150596.

[2]   Mousa, S. A., Shalan, A. E., Hassan, H. H., Ebnawaled, A. A., & Khairy, S. A. (2022). Enhanced the photocatalytic degradation of titanium dioxide nanoparticles synthesized by different plant extracts for wastewater treatment. *Journal of Molecular Structure*, 1250, 131912.

[3]   Rafique, M., Tahir, R., Gillani, S. S. S. A., Tahir, M. B., Shakil, M., Iqbal, T., & Abdellahi, M. O. (2022). Plant-mediated green synthesis of zinc oxide nanoparticles from Syzygium Cumini for seed germination and wastewater purification. *International Journal of Environmental Analytical Chemistry*, 102(1), 23–38.

[4]   Lin, Y., Jin, X., Khan, N. I., Owens, G., & Chen, Z. (2022). Bimetallic Fe/Ni nanoparticles derived from green synthesis for the removal of arsenic (V) in mine wastewater. *Journal of Environmental Management*, 301, 113838.

[5]   Modi, S., Yadav, V. K., Choudhary, N., Alswieleh, A. M., Sharma, A. K., Bhardwaj, A. K., . . . Jeon, B. H. (2022). Onion peel waste mediated-green synthesis of zinc oxide nanoparticles and their phytotoxicity on mung bean and wheat plant growth. *Materials*, 15(7), 2393.

[6]   Iqbal, T., Masood, A., Khalid, N. R., Tahir, M. B., Asiri, A. M., & Alrobei, H. (2022). Green synthesis of novel lanthanum doped copper oxide nanoparticles for photocatalytic application: Correlation between experiment and COMSOL simulation. *Ceramics International*, 48(10), 13420–13430.

[7]   Rather, M. Y., & Sundarapandian, S. (2022). Facile green synthesis of copper oxide nanoparticles and their Rhodamine-b dye adsorption property. *Journal of Cluster Science*, 33(3), 925–933.

[8]   Cao, M., Zhuang, Z., Liu, Y., Zhang, Z., Xuan, J., Zhang, Q., & Wang, W. (2022). Peptide-mediated green synthesis of the MnO2@ ZIF-8 core–shell nanoparticles for

efficient removal of pollutant dyes from wastewater via a synergistic process. *Journal of Colloid and Interface Science*, 608, 2779–2790.

[9]   Ri, K., Han, C., Liang, D., Zhu, S., Gao, Y., & Sun, T. (2022). Nanocrystalline erdite from iron-rich sludge: Green synthesis, characterization and utilization as an efficient adsorbent of hexavalent chromium. *Journal of Colloid and Interface Science*, 608, 1141–1150.

[10]  Rashtbari, Y., Sher, F., Afshin, S., Hamzezadeh, A., Ahmadi, S., Azhar, O., . . . Poureshgh, Y. (2022). Green synthesis of zero-valent iron nanoparticles and loading effect on activated carbon for furfural adsorption. *Chemosphere*, 287, 132114.

[11]  Al-Enazi, N. M. (2022). Optimized synthesis of mono and bimetallic nanoparticles mediated by unicellular algal (diatom) and its efficiency to degrade azo dyes for wastewater treatment. *Chemosphere*, 135068.

[12]  Khan, F., Shahid, A., Zhu, H., Wang, N., Javed, M. R., Ahmad, N., . . . Mehmood, M. A. (2022). Prospects of algae-based green synthesis of nanoparticles for environmental applications. *Chemosphere*, 293, 133571.

[13]  Queiroz, R. N., Prediger, P., & Vieira, M. G. A. (2022). Adsorption of polycyclic aromatic hydrocarbons from wastewater using graphene-based nanomaterials synthesized by conventional chemistry and green synthesis: A critical review. *Journal of Hazardous Materials*, 422, 126904.

[14]  Nasrollahzadeh, M., Sajjadi, M., Iravani, S., & Varma, R. S. (2021). Green-synthesized nanocatalysts and nanomaterials for water treatment: Current challenges and future perspectives. *Journal of Hazardous Materials*, 401, 123401.

[15]  Devatha, C. P., Thalla, A. K., & Katte, S. Y. (2016). Green synthesis of iron nanoparticles using different leaf extracts for treatment of domestic waste water. *Journal of Cleaner Production*, 139, 1425–1435.

[16]  Yazdi, T., Ehsan, M., Amiri, M. S., Akbari, S., Sharifalhoseini, M., Nourbakhsh, F., . . . Es-Haghi, A. (2020). Green synthesis of silver nanoparticles using helichrysum graveolens for biomedical applications and wastewater treatment. *BioNanoScience*, 10(4), 1121–1127.

[17]  Salem, S. S., & Fouda, A. (2021). Green synthesis of metallic nanoparticles and their prospective biotechnological applications: an overview. *Biological Trace Element Research*, 199(1), 344–370.

[18]  Fan, X. M., Yu, H. Y., Wang, D. C., Mao, Z. H., Yao, J., & Tam, K. C. (2019). Facile and green synthesis of carboxylated cellulose nanocrystals as efficient adsorbents in wastewater treatments. *ACS Sustainable Chemistry & Engineering*, 7(21), 18067–18075.

[19]  Devatha, C. P., & Thalla, A. K. (2018). Green synthesis of nanomaterials. In *Synthesis of inorganic nanomaterials* (pp. 169–184). Woodhead Publishing, Cambridge, United Kingdom.

[20]  Goutam, S. P., Saxena, G., Roy, D., Yadav, A. K., & Bharagava, R. N. (2020). Green synthesis of nanoparticles and their applications in water and wastewater treatment. In *Bioremediation of industrial waste for environmental safety* (pp. 349–379). Springer, Singapore.

[21]  Rambabu, K., Bharath, G., Banat, F., & Show, P. L. (2021). Green synthesis of zinc oxide nanoparticles using Phoenix dactylifera waste as bioreductant for effective dye degradation and antibacterial performance in wastewater treatment. *Journal of Hazardous Materials*, 402, 123560.

[22]  Das, C., Sen, S., Singh, T., Ghosh, T., Paul, S. S., Kim, T. W., . . . Biswas, G. (2020). Green synthesis, characterization and application of natural product coated magnetite nanoparticles for wastewater treatment. *Nanomaterials*, 10(8), 1615.

[23]  Das, P., Ghosh, S., Ghosh, R., Dam, S., & Baskey, M. (2018). Madhuca longifolia plant mediated green synthesis of cupric oxide nanoparticles: A promising environmentally

sustainable material for waste water treatment and efficient antibacterial agent. *Journal of Photochemistry and Photobiology B: Biology*, 189, 66–73.

[24]   Luo, X., You, Y., Zhong, M., Zhao, L., Liu, Y., Qiu, R., & Huang, Z. (2022). Green synthesis of manganese–cobalt–tungsten composite oxides for degradation of doxycycline via efficient activation of peroxymonosulfate. *Journal of Hazardous Materials*, 426, 127803.

[25]   Vinayagam, R., Pai, S., Murugesan, G., Varadavenkatesan, T., Kaviyarasu, K., & Selvaraj, R. (2022). Green synthesized hydroxyapatite nanoadsorbent for the adsorptive removal of AB113 dye for environmental applications. *Environmental Research*, 212, 113274.

[26]   Mahmood, K., Amara, U., Siddique, S., Usman, M., Peng, Q., Khalid, M.,...Ashiq, M. N. (2022). Green synthesis of Ag@ CdO nanocomposite and their application towards brilliant green dye degradation from wastewater. *Journal of Nanostructure in Chemistry*, 12(3), 329–341.

# Index

Page numbers in *italics* indicate figures; page numbers in **bold** indicate tables.

## A

actinomycetes, plastic degradation, *19*
activated carbon (AC)
    photocatalytic degradation performance by
        $TiO_2$ doped, 56–57
    synthesis of $TiO_2$/carbon and derivatives,
        54–55
    titanium oxide ($TiO_2$) on, 50
aerobic stirred reactor, membrane bioreactor
    (MBR) with, 121–123
agricultural industry, wastewater from, 101
algae
    bioremediation of dyes, 7, **8**
    bioremediation of microplastics, 14, 18
    bioremediation of organic pollutants,
        13, **14**
    green synthesis of iron oxide nanoparticles
        (IONPs) using biomass, 33
    heavy metal bioremediation, 11
    phytoremediation and, 145
    plastic degradation, *19*
anaerobic stirred reactor, membrane bioreactor
    (MBR) with, 123–124

## B

bacteria
    bioremediation of dyes, 4, **5**
    bioremediation of microplastics, 14, **16–17**
    bioremediation of organic pollutants, 12
    contaminants of water, 97–98
    green synthesis of iron oxide nanoparticles
        (IONPs) using, 33–34
    heavy metal bioremediation, 11, **11**
    microalgal/bacterial mechanism for
        wastewater purification, 104–105, *106*
    plastic degradation, *19*
    studies on bioremediation of dyes, **5**
    up-flow anaerobic sludge blanket reactor for
        wastewater treatment, 105, *107*
    wastewater, 110–112
bacterial cellulose (BNC), 59
ballast water management (BWM), 132
    critical issues for, *133*
    international convention of, 132
    literature review, 134
    offshore and onshore ballast system, *136*
    role in wastewater treatment, 141

    systematic overview of, *136*
    timeline of events for complement of, *134*
ballast water treatment
    actual treatment system on M/V LOCH
        MELFORT, *138*
    basic parameters of LOCH MELFORT
        ship, **138**
    basic parameters of system, **141**
    case study of Western Australia ports,
        137–138, 141
    controlling panel on M/V LOCH
        MELFORT, *140*
    mechanism, *133*
    on-board, system, *135*
    recent works related to, *135*
    solenoid valve, *139*
    "strategy manager" model, 137
    suction/discharge valve system, *139*
    technologies, 134–135
    ultraviolet (UV) treatment method, 137, *137*
battery industry, wastewater from, 99–100
Bhopal gas tragedy, 1
bioaugmentation, 2
biofilters, 2
biological oxygen demand (BOD), 97
    agricultural industry, 101
    food industry, 103–104
    membrane bioreactor (MBR) system, 118
    removal with stirred reactor, 122
biological techniques, wastewater treatment, 96
biomass residues, sources of, *51*
biopile, 3
biopolymers, classification of, based on origin, *52*
bioreactors, 2
bioremediation, 1–3, 21
    advantages and limitations, 21
    dyes, 3–4, 7
    ex situ techniques, 2–3
    heavy metals, 11, **11, 12**
    in situ techniques, 2
    microplastics, 13–14, 18
    organic pollutants, 12–13
    pharmaceutical waste, 20–21
    term, 1
    types of, 2–3
bios, 1
bioslurping, 2
biosparging, 2
bioventing, 2

**C**

cadmium (Cd), **147**
    toxicity of, 149
    *see also* phytoremediation
cellulose acetate (CA), 64
cellulose nano-crystals (CNCs), 64
cellulose nano fibres (CNFs), 64
chemical oxygen demand (COD), 97
    agricultural industry, 101
    food industry, 103–104
    removal with stirred reactor, 123
Chernobyl incident, 1
*Chlorella vulgaris*, heavy metal
    bioremediation, 11
co-axial electrospun fibres, fabrication for water
    filtration, *68*, 69

**D**

dairy industry
    properties of effluents from, **103**
    wastewater from, 103
degradation steps, microbial plastic, *15*
Dorr-Oliver, Inc., 121
DSFF (downflow stationary fixed film) reactor
    basic layout of, *126*
    membrane bioreactor (MBR) with, 125, 127
dyes, 3
    bioremediation of, 3–4, 7
    literature on bioremediation of, using algal
        species, **8**
    literature on bioremediation of, using
        enzymes, **10**
    literature on bioremediation of, using yeast, **9**
    literature on bioremediation using fungal
        species, **6**
    photocatalytic degradation pathways using
        IONPs, *39*
    studies on bioremediation of, using bacteria, **5**
    synthetic, 3, **4**

**E**

earth, distribution of water, 110, *110*
electrical power plant
    wastewater from, 100
    wastewater treatment technologies, *100*
electrospinning coupled with electrospraying,
    fabrication for water filtration, *68*, 70
electrospun membrane fabrication
    application for pressure-driven filtration in
        water treatment, 70–72
    basic layout of process for water treatment, *67*
    basic principle for water treatment, 65, 67
    co-axial electrospinning, *68*, 69

electrospinning coupled with electrospraying,
    *68*, 70
    materials used for, *66*
    melt phase splitting and dispersion, *68*, 70
    tri-axial electrospinning, *68*, 68–69, *69*
    types of, *68*, 68–70
electrospun nanofibre and composites
    application for pressure-driven filtration in
        water treatment, 70–72
    drawbacks of using in wastewater filtration/
        treatment, 72, 74
    microfiltration (MF), 63, 72, *73*
    nanofiltration (NF), 63, 71, *73*
    reverse osmosis (RO), 63, 71
    ultrafiltration (UF), 63, 71–72, *73*
Environmental Protection Agency, 1
enzymes
    bioremediation of dyes, 7, **10**
    bioremediation of microplastics, 18, *19*

**F**

fluidized bed reactor, membrane bioreactor
    (MBR) with, 124–125
food industry, wastewater from, 103–104
Fukushima incident, 1
fungi
    bioremediation of dyes, 4, **6**
    bioremediation of microplastics, 14, **18**
    green synthesis of iron oxide nanoparticles
        using fungal biomass, 31–33
    heavy metal bioremediation, 11, **12**
    plastic degradation, *19*
    studies on bioremediation of dyes, **6**

**G**

green nanotechnology in wastewater treatment,
    79–80, 87–88
    applications, **87**
    filtration technique, 84
    future plans, 86
    heavy metal and ions removal by
        nanoparticles (NPs), 86, *88*
    magnetic NPs (MNPs), 83, *83*
    metal-based nanomaterials, 81–82
    nanocatalyst, 83
    nano membranes, 83–84
    organic compounds removal by NPs, 85–86
    pollutant adsorption by NPs, 84–85
    possible reaction mechanism of
        photocatalytic reaction, *86*
    types of nanomaterials, 80–84, **84**
green synthesis
    nanoparticles (NPs), *155*, 156–158
    protocols, 41–42

green synthesis of iron oxide nanoparticles (IONPs)
  algal biomass for, 33
  bacteria for, 33–34
  fungal biomass for, 31–33
  plant species for, 34

**H**

heavy metals
  bioremediation, 11, **11**, **12**
  cadmium, 149
  iron oxide nanoparticles as adsorbent for,
    34, 36
  mercury, 147–149
  pollution, 7
  presence in environment, **147**
  sources of mercury in surroundings, **148**
  toxicity of, 146–149
  *see also* phytoremediation

**I**

industrial wastes, *see* phytoremediation
international maritime organization (IMO), 132,
    133, *134*
inverse fluidization bed reactor
  layout of, *127*
  membrane bioreactor (MBR) with, 127
iron ox-hydroxide (FeOOH), 41
iron oxide nanoparticles (IONPs), 29, 29–31,
    41–42
  as adsorbent for heavy metals, 34, 36
  as adsorbent for organic contaminants, 37
  adsorption and photocatalytic degradation of
    pollutants using, *38*
  advantages and versatile applications of, *31*
  applications for water treatment, 34, 36–38
  biological synthesis of, 30–31
  biosynthesis of, *32*, *35*
  challenges and future perspective of using,
    38–41
  forms of, *30*
  green synthesis of, using algal biomass, 33
  green synthesis of, using bacteria, 33–34
  green synthesis of, using fungal biomass,
    31–33
  green synthesis of, using plant species, 34
  methods for synthesis of, *29*
  "nanozymes," 30
  as photo-catalyst, 37–38
  protocol for synthesis of, using green routes, *36*

**L**

land farming, 2
leather industry, wastewater from, 101

**M**

macrophytes, phytoremediation and, 144–145
magnetic nanoparticles (MNPs), 83, *83*, 154
magnetic particle imaging (MPI), 29
magnetic resonance imaging (MRI), 29
melt phase splitting and dispersion, fabrication
    for water filtration, *68*, 70
membrane bioreactor (MBR) system, 117–120, 128
  basic layout of, *120*
  basic layout of external and submerged, *119*
  history of development of, 120–121
  MBR with aerobic stirred reactor, 121–123
  MBR with anaerobic stirred reactor, 123–124
  MBR with DSFF reactor, 125, *126*, 127
  MBR with fluidized bed reactor, 124–125
  MBR with inverse fluidized bed reactor,
    127, *127*
  MBR with semi-fluidized bed reactor, 125, *126*
  process of, *118*
  types of, 121–127
  types of configuration/design in, 121
mercury (Hg), **147**
  sources in surroundings, **148**
  toxicity of, 147–149
  *see also* phytoremediation
metal nanoparticles, manufacture of, 32
metal oxides (MO), 28–29
  nanoparticles (NPs), 80
  wastewater remediation applications, 38–39
microalgae
  contaminants of water, 98
  microalgal/bacterial mechanism for
    wastewater purification, 104–105, *106*
  wastewater, 110–112
microalgal/bacterial mechanism
  advantages and disadvantages of, for
    wastewater, 105, 107
  purification of wastewater, 104–105, *106*
  simplified layout of synergistic
    interaction, *109*
microcrystalline cellulose (MFC), 59
microfiltration (MF)
  membrane filtration for water treatment, 72, *73*
  pressure-driving membranes, 63
microplastics (MPs)
  bioremediation, 13–14, 18
  degradation steps, *15*
  pollution, 13
  role of enzymes in bioremediation, 18, *19*
mines industry, wastewater from, 102–103

**N**

nanocatalyst, 83
nanocrystalline cellulose (NCC), 59

nanofiltration (NF)
    membrane filtration for water treatment, 71, *73*
    pressure-driving membranes, 63
nanomaterials, metal-based, 81–82
nano membranes, 83–84
nanoparticles (NPs), 28–29
    characterisation of, *156*, 158
    characteristics of, **155**
    filtration technique, 84
    green, 79–80
    green synthesis of, *155*, 156–158
    heavy metal and ions removal by, 86, **87**, *88*
    iron oxide NPs (IONPs), 29
    metal oxide NPs, 80
    organic compound removal by, 85–86
    pollutant adsorption by, 84–85
    synthesis and wastewater treatment,
        153–154, 156
    wastewater treatment by, **87**, *154*
    wastewater treatment by green NPs, 84–86
nanotechnology
    field of, 28
    wastewater treatment by, *81*
    *see also* green nanotechnology in wastewater
        treatment
nanozymes, iron oxide nanoparticles
    (IONPs) as, 30
*Nostoc* sp., phytoremediation and, 145–146
nuclear power plant, wastewater from, 100

**O**

oil mill industry, wastewater from, 102
organic chemical production industries,
    wastewater from, 102
organic compounds
    bio-remediation of, 12–13
    iron oxide nanoparticles as adsorbent for,
        contaminants, 37
    photocatalytic activity for degradation of, *49*
    photocatalytic degradation, 47, 49
    pollutants, 11–12

**P**

petroleum and petro-chemical industries,
    wastewater from, 102
pharmaceutical waste
    bioremediation of, 20–21
    industrial wastewater, 102
    occurrence in effluent water from
        wastewater, **20**
photo-catalyst, iron oxide nanoparticles (IONPs)
    as, 37–38, *38*
phytoremediation, 144, 149
    cadmium, 149
    macrophytes, 144–145
    mercury, 147–149

*Nostoc* sp., 145–146
*Scenedesmus* sp., 145
    toxicity of heavy metals, 146–149
plant extracts, 80
plant species, green synthesis of iron oxide
    nanoparticles (IONPs) using, 34
pollution
    heavy metals, 7
    microplastics, 13
    organic compounds, 11–12
polyethersulfone (PES), 64
polyvinylidene fluoride (PVDF), 64
Port State Control (PSC), 136
pulp and paper industry, wastewater from, 101

**R**

remediate, 1
Rensselaer Polytechnic Institute, 121
reverse osmosis (RO)
    membrane filtration for water treatment, 71
    pressure-driving membranes, 63
    water filtration membranes, 64

**S**

*Scenedesmus* sp., phytoremediation and, 145
sea-ocean environment, *see* ballast water
    management (BWM)
semi-fluidized bed reactor
    layout of, *126*
    membrane bioreactor (MBR) with, 125
sewage water
    standard treatment, 110
    treatment by nanoparticles, *154*
State Inspectorate for the Protection of the
    Environment (PIOS), 97
steel and iron industry
    toxicity impact of, **104**
    wastewater from, 103
Swedish Environmental Protection Agency, 108
synthetic dyes, classification of, 3, **4**

**T**

textile industry, wastewater from, 101
$TiO_2$-based activated carbon (AC), 59
    basic layout of photocatalytic reactor
        containing, as water treatment
        membrane, *56*
    mechanism of photo-degradation of organic
        pollutant using, *57*
    photocatalytic degradation performance by,
        56–57
    synthesis of, and its derivatives, 54–55
$TiO_2$ doped biopolymer composites
    challenges and future perspectives of, 58
    water purification using, 58–59

TiO$_2$ doped carbonaceous substrate, photocatalytic degradation performance, 56–57
TiO$_2$-doped cellulose
  advanced oxidation process for water treatment using, 47, *48*
  cellulose derived carbon as support matrix of TiO$_2$, 52
  nucleation and growth of TiO$_2$ nanoparticles over cellulosic filament, *53, 54*
  photocatalytic degradation performance by, for water treatment, 55–56
  porous TiO$_2$-based composites using cellulosic filament, 52, 54
  synthesis of, and its derivatives, 51–54
  TiO$_2$ coating over cellulosic filaments, 51–52
  utilization of biopolymers for stabilising nanoparticles, 50
tri-axial electrospun fibres, fabrication for water filtration, *68*, 68–69

**U**

ultrafiltration (UF)
  membrane filtration for water treatment, 71–72, *73*
  pressure-driving membranes, 63
ultraviolet (UV) system, ballast water treatment, 137, *137*

**W**

wastewater
  bacteria and, 110–112
  biological remediation, 117
  formation of, *155*
  microalgae and, 110–112
wastewater purification
  synthesis of TiO$_2$/carbon and its derivatives for, 54–55
  synthesis of TiO$_2$/cellulose and its derivatives for, 51–54
wastewater remediation, iron oxide nanoparticles (IONPs) as, 38–41
wastewater treatment
  advantages and disadvantages of hybrid microalgal/bacterial approach, 105, 107–108
  agricultural industry, 101
  battery industry, 99–100
  constituents of, 97, *97*
  contaminants in, 97–98
  dairy industry, 103, **103**
  electrical power plant, 100, *100*
  food industry, 103–104

  hybrid mechanism of microalgal/bacterial approach, 104–105, *106*
  implications of, 99
  leather industry, 101
  magnetic nanoparticles (MNPs), 83, *83*
  metal-based nanomaterials, 81–82
  mines industry, 102–103
  nanocatalyst, 83
  nano membranes, 83–84
  nanoparticle synthesis and, 153–154, *154*, 156
  nuclear power plant, 100
  oil mill industry, 102
  organic chemical production industries, 102
  petroleum and petro-chemical industries, 102
  pharmaceutical industry, 102
  pulp and paper industry, 101
  steel and iron industry, 103, **104**
  synergistic interaction for hybrid bacterial/algal system, *109*
  technology, 135–137
  textile industry, 101
  types of nanomaterials in, 80–84
  up-flow anaerobic sludge blanket reactor using bacterial species, 105, *107*
  *see also* green nanotechnology in wastewater treatment
water, distribution of earth's, 110, *110*
water treatment
  advanced oxidation process using TiO$_2$ doped cellulose/carbon matrix, *48*
  basic layout of photocatalytic reactor containing TiO$_2$-based ACs as membrane, *56*
  biological techniques, *96*
  chemical techniques, *96*
  photocatalytic degradation performance by TiO$_2$ doped carbonaceous substrate, 56–57
  photocatalytic degradation performance by TiO$_2$ doped cellulosic, 55–56
  physical techniques, *96*
  types of technology, 95, *96*
  *see also* electrospun membrane treatment
Western Australia ports, case study, 137–138, 141
windrows, 3
World Health Organization, 110

**Y**

Yeast, bioremediation of dyes, 7, **9**

**Z**

zero valent iron (ZVI) nanoparticles, 41
zinc oxide (ZnO) nanoparticles, 65

Taylor & Francis Group
an **informa** business

# Taylor & Francis eBooks

www.taylorfrancis.com

A single destination for eBooks from Taylor & Francis
with increased functionality and an improved user
experience to meet the needs of our customers.

90,000+ eBooks of award-winning academic content in
Humanities, Social Science, Science, Technology, Engineering,
and Medical written by a global network of editors and authors.

## TAYLOR & FRANCIS EBOOKS OFFERS:

A streamlined
experience for
our library
customers

A single point
of discovery
for all of our
eBook content

Improved
search and
discovery of
content at both
book and
chapter level

## REQUEST A FREE TRIAL
support@taylorfrancis.com

Routledge
Taylor & Francis Group

CRC Press
Taylor & Francis Group

For Product Safety Concerns and Information please contact our EU
representative GPSR@taylorandfrancis.com
Taylor & Francis Verlag GmbH, Kaufingerstraße 24, 80331 München, Germany

www.ingramcontent.com/pod-product-compliance
Lightning Source LLC
Chambersburg PA
CBHW070722220326
41598CB00024BA/3263